Spiral Structure in Galaxies

Spiral Structure in Galaxies

Marc S Seigar
Department of Physics and Astronomy, University of Minnesota Duluth, USA

Morgan & Claypool Publishers

Rights & Permissions
To obtain permission to re-use copyrighted material from Morgan & Claypool Publishers, please contact info@morganclaypool.com.

ISBN   978-1-6817-4609-8 (ebook)
ISBN   978-1-6817-4608-1 (print)
ISBN   978-1-6817-4611-1 (mobi)

DOI   10.1088/978-1-6817-4609-8

Version: 20170601

IOP Concise Physics
ISSN 2053-2571 (online)
ISSN 2054-7307 (print)

A Morgan & Claypool publication as part of IOP Concise Physics
Published by Morgan & Claypool Publishers, 40 Oak Drive, San Rafael, CA, 94903 USA

IOP Publishing, Temple Circus, Temple Way, Bristol BS1 6HG, UK

For my wife, Colleen,

and my boys, David and Andrew

Contents

Preface

A spiral galaxy is a type of galaxy originally described by Edwin Hubble in his 1936 work *The Realm of Nebulae* and, as such, forms part of the Hubble sequence. Spiral galaxies consist of a flat, rotating disk containing stars, gas and dust, and a central concentration of stars known as the bulge. These are surrounded by a much fainter halo of stars, many of which reside in globular clusters.

Spiral galaxies are named for the spiral structures that extend from the center into the galactic disk. The spiral arms are sites of ongoing star formation and are brighter than the surrounding disk because of the young, hot OB stars that inhabit them. Roughly two-thirds of all spirals are observed to have an additional component in the form of a bar-like structure, extending from the central bulge, at the ends of which the spiral arms begin. The proportion of barred spirals relative to their barless cousins has changed over the history of the Universe, with only about 10% containing bars about 8 billion years ago, to roughly a quarter 2.5 billion years ago, until the present, where over two-thirds of the galaxies in the visible Universe have bars.

This book focuses on why these disk-shaped (or spiral) galaxies have spiral arms. Why do these structures exist? Why are they so stable? And what is the connection between the spiral arms and the star formation that is seen within them? In this book you will find the answers to these questions and more.

Marc S Seigar
Department of Physics and Astronomy
University of Minnesota Duluth
Duluth, MN, USA
5 May 2017

Acknowledgements

I wish to thank all of my family and friends. In particular, I want to thank my PhD advisor, Professor Philip James, without whom I would never have become interested in spiral structure in galaxies. I wish to thank all of my collaborators, in particular Andy Adamson, Misty Bentz, Joel Berrier, Jane Buckle, James Bullock, Benjamin Davis, Chris Davis, Herwig Dejonghe, Tim de Zeeuw, Fabio Gastaldello, Alister Graham, Philip Humphrey, Helmut Jerjen, Daniel Kennefick, Julia Kennefick, Olga Kuhn, Claud Lacy, Sandy Leggett, Paul Lynam, Burcin Mutlu-Pakdil, Ivanio Puerari, Nickolas Rees, Heikki Salo, Doug Shields, Amber Sierra (Harrington), Massimo Stiavelli, Patrick Treuthardt, Thor Wold, Watson Varricatt, and Luca Zappacosta, and many more.

I want to thank all of my friends and colleagues who have helped me through the years, particularly Haydar Al-Shukri, Matt Andrews, Brian Berry, John Bush, Chris Collins, Toni Empl, Jeff Gaffney, Michael Gealt, Anindya Ghosh, Joshua Hamilton, Molly Harney, Keith Hudson, Darin Jones, Tansel Karabacek, Johanna Lewis, Howard Mooers, Patrick Pellicane, Julian Post, Jeff Robertson, Jim Rock, Derek Sears, Hazel Sears, and Amber Straughn.

My friends with whom I have had so much fun over the years, I thank you all: Liz Alvarez, Roberto Avila, Renee and Kirt Booth, Alex Bouquin, Marc Cohen, Richard Dolman, Michael Fahrenwald, Paul Harman, Breanna and Bill Johnson, Martin Johnston, Melissa and Ivan Mitchell, Erik Rau and Elle Bublitz, Helene Schuller and Todd Chelson, Jamie and Dan Sweeney, Sonal and Gavin Whitlock, amongst others.

I wish to thank the editorial and production teams and Morgan & Claypool and IOP Publishing for making the process of writing and publishing this book fairly easy and straightforward.

Finally, I would like to thank my parents, Vivienne and Michael Seigar, who always encouraged me. Without them, I would not be where I am today.

Author Biography

Marc S Seigar

Marc S Seigar is a Professor of Astrophysics and Associate Dean of the Swenson College of Science and Engineering at the University of Minnesota Duluth (UMD). He is also the current director of the Marshall W Alworth Planetarium at UMD. Prior to his arrival at UMD, he worked as a Professor of Astrophysics at the University of Arkansas at Little Rock, a Project Scientist at the University of California, Irvine, and a Staff Astronomer at the United Kingdom Infrared Telescope. Professor Seigar has published numerous papers and conference proceedings articles in the field of galaxy dynamics, spiral structure, and dark matter.

Chapter 1

The discovery of spiral galaxies

William Parsons, 3rd Earl of Rosse (a photograph of whom can be seen in figure 1.1) was an Anglo-Irish astronomer who had several telescopes built. His 72 inch telescope, built in 1845 (also shown in figure 1.1) and colloquially known as the 'Leviathan of Parsonstown', was the world's largest telescope, in terms of aperture size, until the early 20th century.

During the 1840s, Rosse had the Leviathan of Parsonstown built, a 72 inch (6 feet/1.83 m) telescope at Birr Castle, Parsonstown, County Offaly, Ireland. The 72 inch (1.8 m) telescope replaced a 36 inch (910 mm) telescope that he had built previously. He had to invent many of the techniques he used for constructing the Leviathan, both because its size was without precedent and because earlier telescope builders had guarded their secrets or had simply failed to publish their methods. Details of the metal, casting, grinding and polishing of the 3 ton 'speculum' were presented in 1844 at the Belfast Natural History Society. Rosse's telescope was considered a marvelous technical and architectural achievement, and images of it were circulated widely within the British Commonwealth. Building of the Leviathan began in 1842 and it was first used in 1845; regular use waited another two years, due to the Great Irish Famine. Using this telescope, Rosse saw and cataloged a large number of nebulae (including a number that would later be recognized as galaxies).

Lord Rosse performed astronomical studies and discovered the spiral nature of some nebulae, today known to be spiral galaxies. Rosse's telescope Leviathan was the first to reveal the spiral structure of M51, a Galaxy nicknamed later as the 'Whirlpool Galaxy', and his drawings of it (see figure 1.2) closely resemble modern photographs.

At the time of Rosse's observations of M51 (in 1845), it was referred to as a spiral nebula. It was the first of many spiral nebulae that he observed with his Leviathan. In 1845, there was no method for determining the distances to spiral nebulae, and so it was believed that they were part of the Milky Way Galaxy. It would take an

doi:10.1088/978-1-6817-4609-8ch1

Figure 1.1. Left panel: William Parsons, 3rd Earl of Rosse, image courtesy Wikipedia. Right panel: The 72 inch refracting telescope at Birr Castle, the largest telescope of its time, colloquially known as the Leviathan of Parsonstown'.

Figure 1.2. Lord Rosse's drawing of M51, the 'Whirlpool Galaxy', as observed through the Leviathan of Parsonstown.

analysis of a certain type of star, known as Cepheid Variables, to provide the first technique to measure the distances to these nebulae.

On 10 September, 1784, Edward Pigott detected the variability of Eta Aquilae, the first known representative of the class of classical Cepheid variables. However, the archetypal star for classical Cepheids is Delta Cephei, discovered to be variable by John Goodricke a few months later. Little was it known at the time that Cepheid variables would hold a key to dramatically changing our view of the Universe in the early part of the 20th century. Fundamental to this would be a discovery by one of the unsung heroes of 20th century astronomy.

Henrietta Swan Leavitt (pictured in figure 1.3) was an American astronomer. A graduate of Radcliffe College, Leavitt started working at the Harvard College Observatory as a 'computer' in 1893, examining photographic plates in order to measure and catalog the brightness of the stars. The term 'computer', in use from the early 17th century meant 'one who computes': a person performing mathematical calculations, before electronic computers became commercially available. Teams of people were frequently used to undertake long and often tedious calculations; the work was divided so that this could be done in parallel. Harvard College Observatory employed a team of women to perform these calculations, and Henrietta Swan Leavitt was one of them. She was hired at the observatory by Edward Charles Pickering and was assigned the task of cataloging stars (in the early 1900s, women were not allowed to operate telescopes). Paid at a rate of just 30 cents per hour, she was reportedly hardworking, serious-minded, and devoted to her family, her church, and her career.

Pickering assigned Leavitt to study 'variable stars', whose luminosity varies over time. According to science writer Jeremy Bernstein, 'variable stars had been of interest for years, but when she was studying those plates, I doubt Pickering

Figure 1.3. Henrietta Swan Leavitt, unknown source.

thought she would make a significant discovery—one that would eventually change astronomy'. Leavitt noted thousands of variable stars in images of the Magellanic Clouds. In 1908, she published her results in the Annals of the Astronomical Observatory of Harvard College, noting that a few of the variables showed a pattern: brighter ones appeared to have longer periods. After further study, she confirmed in 1912 that the Cepheid variables[1] with greater intrinsic luminosity did have longer periods, and that the relationship was quite close and predictable.

Leavitt used the simplifying assumption that all of the Cepheids within each Magellanic Cloud were at approximately the same distances from Earth, so that their intrinsic brightness could be deduced from their apparent brightness (as measured from the photographic plates) and from the distance to each of the clouds. She noted that 'since the variables are probably at nearly the same distance from the Earth, their periods are apparently associated with their actual emission of light, as determined by their mass, density, and surface brightness'.

Her discovery, which she produced from studying some 1777 variable stars recorded on Harvard's photographic plates, is known as the 'period–luminosity relationship' for Cepheid variables: the logarithm of the period is linearly related to the logarithm of the star's average intrinsic optical luminosity. The period–luminosity relationship for Cepheids made them the first 'standard candle' in astronomy, allowing scientists to compute the distances to objects that are too distant for stellar parallax observations. Leavitt was not recognized for her work until after her death.

Edwin Powell Hubble (pictured in figure 1.4) was an American astronomer. He played a crucial role in establishing the fields of extragalactic astronomy and observational cosmology and is regarded as one of the most important astron-omers of all time. Hubble's arrival at Mount Wilson Observatory, California in 1919, coincided roughly with the completion of the 100 inch (2.5 m) Hooker Telescope, then the world's largest telescope. At that time, the prevailing view of the cosmos was that the Universe consisted entirely of the Milky Way Galaxy. Using the Hooker Telescope at Mount Wilson, Hubble identified Cepheid variables in several spiral nebulae, including the Andromeda nebula and the Triangulum nebula. His observations, made in 1922–23, proved conclusively that these nebulae were much too distant to be part of the Milky Way and were, in fact, entire galaxies outside our own, suspected by researchers at least as early as 1755 when Immanuel Kant's *General History of Nature and Theory of the Heavens* appeared. This idea had been opposed by many in the astronomy establishment of the time, in particular by the Harvard University-based Harlow Shapley. Despite the opposition, Hubble, then a 35-year-old scientist, had his findings first published in *The New York Times* on 23 November, 1924, and then more formally presented in the form of a paper at the 1 January, 1925 meeting of the American Astronomical Society.

---

[1] https://en.wikipedia.org/wiki/Cepheid_variable

Figure 1.4. Edwin Hubble observing at the 100 inch telescope at Mount Wilson Observatory, California. Photograph by Margaret Bourke-White, 1937.

Hubble's findings fundamentally changed the scientific view of the Universe. Supporters state that Hubble's discovery of nebulae, outside of our Galaxy, helped pave the way for future astronomers. Although some of his more renowned colleagues simply scoffed at his results, Hubble ended up publishing his findings on nebulae. The so-called nebulae, are, of course, now referred to as galaxies. Hubble's published work earned him an award titled the American Association Prize and five hundred dollars from Burton E Livingston of the Committee on Awards. At the time, the Nobel Prize in Physics did not recognize work done in astronomy, otherwise Hubble would have surely won it. Hubble spent much of the later part of his career attempting to have astronomy considered an area of physics, instead of being its own science. He did this largely so that astronomers—including himself—could be recognized by the Nobel Prize Committee for their valuable contributions to astrophysics. This campaign was unsuccessful in Hubble's lifetime, but shortly after his death, the Nobel Prize Committee decided that astronomical work would be eligible for the physics prize.

This story brings us to the point where we now understand that spiral galaxies exist outside of our own Milky Way. But why are they disk shaped? Any why do they have spiral arms? Why are the arms sustainable for billions of years? These are the questions that will be addressed by this book, starting with how galaxies in the Universe are classified.

Suggested further reading

Hubble E 1926 Extragalactic nebulae *Astrophys. J.* **64** 321–69

Johnson G 2005 *Miss Leavitt's Stars : The Untold Story of the Woman Who Discovered How To Measure the Universe* 1st edn (New York: Norton)

Knight C (ed) 1867 *William Parsons, 3rd Earl of Rosse. Biography: Or Third Division of 'The English Cyclopedia'* vol 5 (London: Bradbury: Evans, & Co.) pp 166–7

Lamb G M 2005 Before computers there were these humans... *Christian Science Monitor* July 5, 2005

Leavitt, Henrietta Swan, Pickering and Edward C Harvard College Observatory Circular **173** 1–3

Lord Rosse's Telescope, The Practical Mechanic and Engineer's Magazine, Feb 1844, p 185

Parsons M 6th Earl of Rosse 1968 William Parsons, third Earl of Rosse *Hermathena* **107** 5–13

Telescopes: Lord Rosse's Reflectors http://amazingspace.org/resources/explorations//groundup/lesson/scopes/rosse/index.php

1912: Henrietta Leavitt Discovers the Distance Key *Everyday Cosmology* N.p., n.d. Web. 20 Oct.2014

Chapter 2

The classification of galaxies

After Edwin Hubble discovered that galaxies outside of the Milky Way existed, he spent some time classifying galaxies based upon their shapes. He came up with a system known as the Hubble sequence or Hubble Tuning Fork Diagram (see figure 2.1). The Hubble sequence is simply a morphological classification scheme for galaxies. The sequence is broken down into two broad categories of galaxies, elliptical galaxies on the left and spiral galaxies on the right. Spirals are further broken down into normal spiral galaxies on the top and barred spirals on the bottom. Lenticular galaxies (or S0 galaxies) are located at the junction between ellipticals and spirals. About 20 percent of galaxies in the Universe do not fall into this scheme, and these are referred to as irregular galaxies.

2.1 Elliptical galaxies

Elliptical galaxies, some examples of which are shown in figure 2.2, are found on the left side of the Hubble sequence. They are classified from E0, which are perfectly circular, to E7, which are highly elongated. They are classified by the letter E followed by a number n. The value of n is an integer determined by the ratio of the semi-minor axis, b, and the semi-major axis, a, of the galaxy using the following equation:

$$n = 10 \times \left(1 - \frac{b}{a}\right) \qquad (2.1)$$

Unlike spiral galaxies, elliptical galaxies are not supported by rotation. The orbits of the constituent stars are random and often very elongated, leading to a shape for the galaxy determined by the speed of the stars in each direction. Faster moving stars can travel further before they are turned back by gravity, resulting in the creation of the long axis[1] of the elliptical galaxy in the direction these stars are moving.

---

[1] http://astronomy.swin.edu.au/cosmos/A/Axis

doi:10.1088/978-1-6817-4609-8ch2

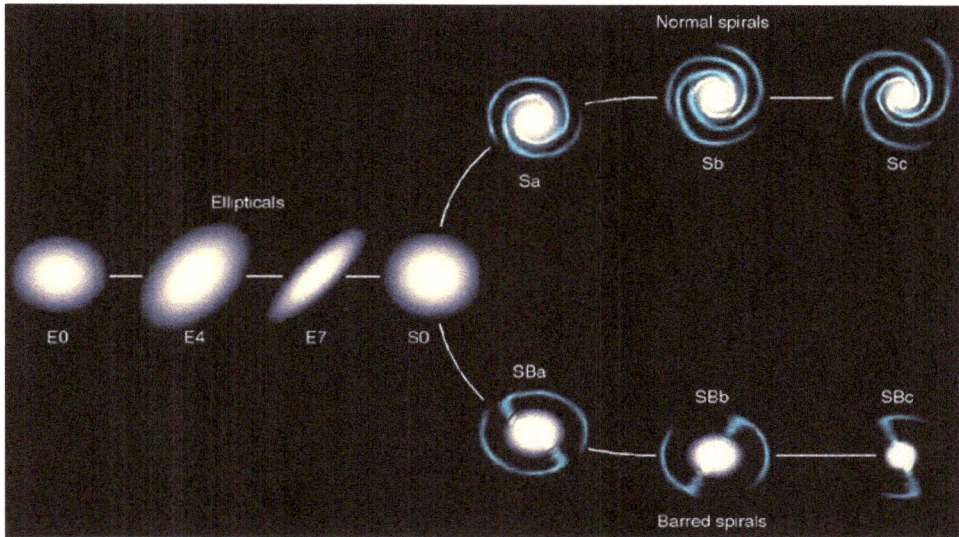

Figure 2.1. The Hubble Tuning Fork Diagram, which classifies galaxies into two categories, elliptical galaxies on the left and spiral galaxies on the right. Spiral galaxies are further subdivided into two categories, normal spirals on the top and barred spirals on the bottom. A third class of galaxy, S0s or lenticular galaxies is placed at the junction between ellipticals and spirals.

Figure 2.2. Left panel: NGC 5128 (or Centaurus A) is a type E0 peculiar elliptical galaxy. Image courtesy of NOAO/AURA/NSF. Right panel: the giant elliptical galaxy ESO 325-G004. Image courtesy of NASA, ESA, and the Hubble Heritage Team.

Elliptical galaxies are the most common type of galaxy in the Universe. They also span the widest range in mass, from dwarf elliptical galaxies containing about 10^7 solar masses in stars to giant elliptical galaxies containing about 10^{13} solar masses in stars. In general, they are red in color, which signifies that they are no longer forming stars. Astronomers therefore refer to them as 'red and dead'.

2.2 Lenticular galaxies

Lenticular galaxies are found at the junction between elliptical and spiral galaxies. They have a bulge surrounded by a stellar disk, just like spiral galaxies, but they have no spiral structure. In some cases they contain central bars. Elliptical and lenticular galaxies together are referred to as early-type galaxies. An example of a lenticular galaxy is shown in figure 2.3.

2.3 Spiral galaxies

Spiral galaxies (see figure 2.4) are seen on the right hand side of the Hubble tuning fork. They are subdivided into two broad classes. On the top arm of the tuning fork, we see normal spiral galaxies. They are classified by use of the letter S. On the

Figure 2.3. The lenticular galaxy PGC 836677, imaged as part of the Coma Cluster Survey. Image courtesy of NASA/ESA/STScI.

Figure 2.4. Left panel: the Pinwheel Galaxy (also known as Messier 101 or NGC 5457) is a normal spiral galaxy. Image credit: European Space Agency and NASA. Right panel: NGC 1300, a barred spiral galaxy in the Southern hemisphere. Image credit: Anglo-Australian Observatory.

bottom, we see barred spiral galaxies, which are classified by use of the letters SB. For both normal and spiral galaxies, the classification S or SB is followed by a lower case a, b, or c. The morphological sequences running from Sa to Sc and from SBa to SBc conventionally denote sequences of normal and barred spirals with arms that are increasingly loosely wound, and bulges that become less dominant (smaller in size) with respect to the disk. Also, the fraction of the gas content appears to increase somewhat as one progresses from Sa (or SBa) to later Hubble types.

2.4 The de Vaucouleurs Classification scheme

A more detailed scheme than the Hubble sequence is the de Vaucouleurs Classification scheme (see figure 2.5). This deals with the continuous gradation of bar, ring, and spiral structures. For Hubble types S and SB, de Vaucouleurs adopted SA and SB, and to the subdivisions, a, b and c, de Vaucouleurs added d and m (for Magellanic Irregulars) to denote a transition from well-developed spiral arms to more chaotic structures. Smoother transitions between barred and non-barred structures were also adopted in the de Vaucouleurs scheme, with the SAB classification for oval distortions. Furthermore, types ab, bc, cd, and dm were adopted to denote smoother transitions from tightly wound to loosely wound spiral structures. The labels (r) and (s) were also added to indicate the prominence of ring and/or spiral features. Thus, what Hubble may have classified as Sc, de Vaucouleurs may have reclassified as SA(rs)cd.

Spiral galaxies have a disk-like structure, with a central, spheroidal bulge. They tend to have a low value for the ratio of random velocities to rotational velocities, and the resultant angular momentum accounts for the disk-like structure. At the center of spiral galaxies, the random velocities become more dominant and this accounts for the central bulge.

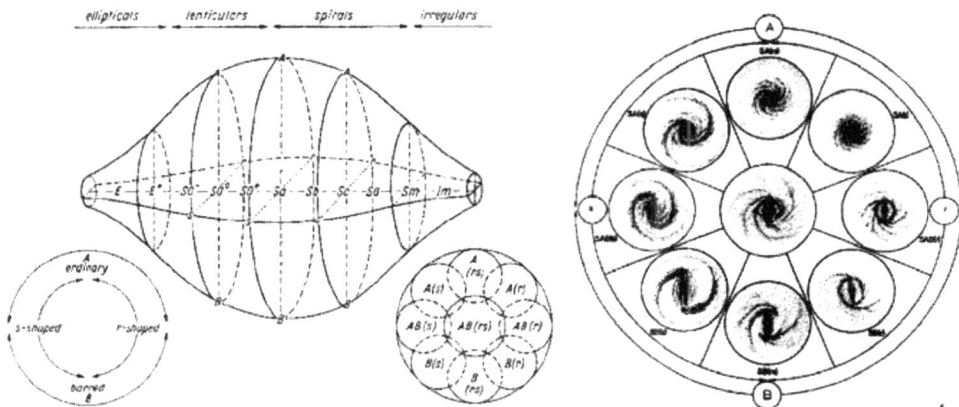

Figure 2.5. Left panel: the de Vaucouleurs classification scheme introduces the notion of a 'classification volume' rather than a one-dimensional sequence. The classification volume is depicted in this sketch most recently published by Buta and Combes (1996) and reproduced by permission of Ron Buta. Right panel: an artist's conception of how the SA/SB and r/s classes interact at stage b (from the left panel), also from Buta and Combes (1996). A typical classification in the de Vaucouleurs system looks like SAB(rs)bc, which is in fact the classification for the Milky Way Galaxy.

The Hubble and de Vaucouleurs classification schemes are possibly determined by three parameters of spiral galaxies. These are as follows:

1. The bulge-to-disk ratio, i.e., the relative distribution of mass between the bulge and disk. Earlier-type spiral galaxies (Sa) seem to have a higher concentration of mass in their bulges than later-type galaxies (Sc). However, studies have shown that only a weak correlation exists between bulge-to-disk ratio and Hubble type (de Jong 1996; Seigar and James 1998a).
2. The spiral arm pitch angle, which is a measure of how tightly wound spiral arms are in galaxies. However, the correlation between pitch angle and Hubble type was also found to be weak (Seigar and James 1998b).
3. Cold gas content, which increases slightly as one progresses from earlier-type spirals to later-type spiral galaxies.

Spiral galaxies are also believed to be embedded in massive dark matter halos. The primary evidence for this comes from the observed rotation curves of spiral galaxies (see figure 2.6). If the visible matter was the only matter in spiral galaxies,

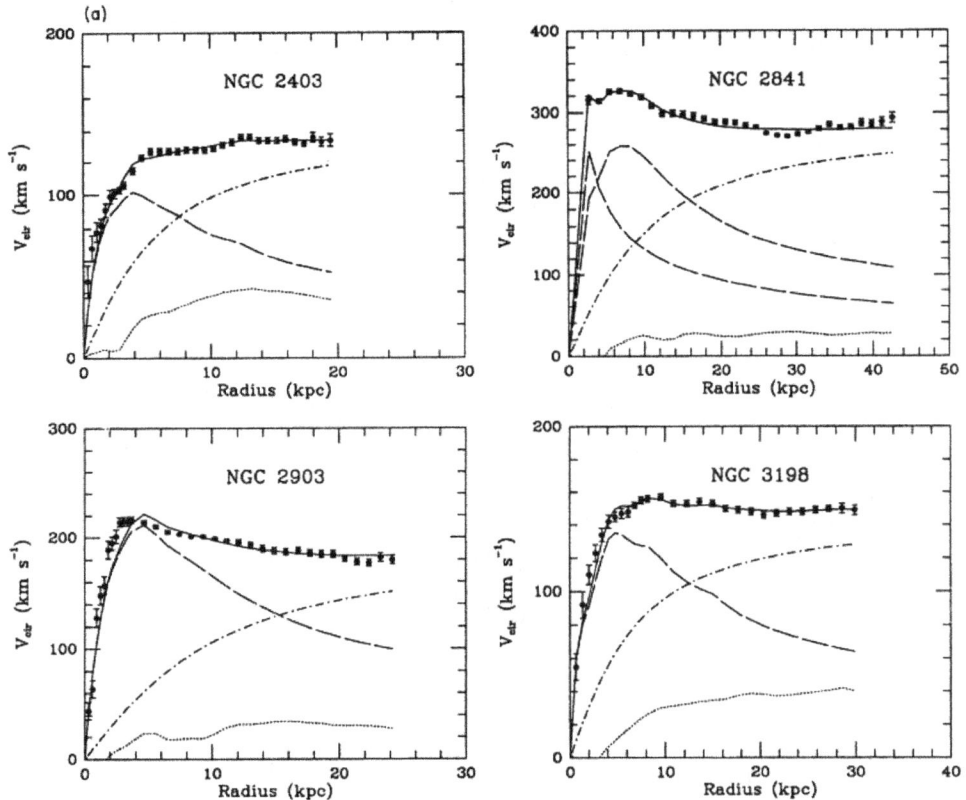

Figure 2.6. Rotation curves of several spiral galaxies, with the contribution from luminous components (dashed), gas (dotted) and dark matter halo (dash-dot). The square blocks are data. The solid line is dark matter model fitting. Data and model fits taken from Begeman *et al* (1991). Reproduced from Bergeman *et al* (1991) with permission of the Royal Astronomical Society.

their rotation curves would turn over and fall off in an approximately Keplerian manner. However, this is not what is observed, but instead rotation curves flatten off and remain flat out to large radii. This applies that approximate 90 percent of mass in spiral galaxies is in the form of dark matter.

2.5 Spiral galaxy bulges

The light profiles of spiral galaxy bulges were originally found to be well parameterized by a $r^{1/4}$ law by Gerard de Vaucouleurs in 1953, i.e.,

$$\mu(r) = \mu_e e^{-7.67[(r/r_e)^{1/4}-1]} \tag{2.2}$$

where $\mu(r)$ is the surface brightness as a function of radius, r_e is the effective radius which contains half of the total light coming from the bulge, and μ_e is the surface brightness at that radius. However, in the last 20 years or so, it has become evident that some spiral galaxy bulges are better fit by exponential light profiles. Indeed, in 1963, astronomer José Luis Sérsic published an alternative light profile, which is now in much wider use for spiral galaxy bulges:

$$\mu(r) = \mu_e e^{-b_n[(r/r_e)^{1/n}-1]} \tag{2.3}$$

The factor b_n is a function of the Sérsic index, n, such that $\Gamma(2n) = 2\gamma(2n, b_n)$, where Γ is the gamma function and γ is the incomplete gamma function (see Graham and Driver 2005). The parameter b_n can be well approximated by $b_n \approx 1.9992n - 0.3271$ for Sérsic indices in the range $1 \leqslant n \leqslant 10$. In the case where $n = 4$, the Sérsic model is equivalent to the $r^{1/4}$ model, and when $n = 1$, it is equivalent to an exponential law.

2.6 Spiral galaxy disks

Disks of spiral galaxies are generally well described by an exponential light profile as follows:

$$\mu(r) = \mu_0 e^{-r/h} \tag{2.4}$$

where μ_0 is the disk central surface brightness and h is the disk scalelength. The exponential form of the light profiles of disks has been explained by various authors using viscous dissipation models.

2.7 Bars in spiral galaxies

About two-thirds of spiral galaxies contain bars. Furthermore, 80 percent of spiral galaxies contain either a bar or an oval distortion. These structures appear to be long-lived because bars have been observed in galaxies at a redshift of $z \approx 1$, which is equivalent to a lookback time of about 8 billion light-years. Indeed, several simulations have shown that bars seem to arise as a natural instability in a cold self-gravitating disk of stars.

There are several ways to parameterize bars. In 1966, Ken Freeman formulated an analytical light profile for a bar as follows:

$$\mu_{bar}(x, y) = \mu_{0,bar}(1 - (x/a_{bar})^2 - (y/b_{bar})^2)^{1/2} \tag{2.5}$$

where $\mu_{0,bar}$ is the bar central surface brightness and a_{bar} and b_{bar} are the semi-major and semi-minor axes of the bar, respectively. However, it has been shown that bar light profiles are described much better by a Ferrers profile as follows:

$$\mu_{bar}(r) = \mu_{0,bar}[1 - (r/r_{out})^2]^{m_{bar}+0.5} \tag{2.6}$$

where m_{bar} is a parameter that defines the shape of bar profiles. The function is defined out to the bar radius, r_{out}. Beyond this radius, the surface brightness of the profile is set to zero. Most bars are best represented with a fit that provides $m_{bar} \leqslant 0.5$.

2.8 Star formation in bars

Patterns seen in spiral galaxies rotate at different speeds when compared to the material. This allows resonances to be set up between the pattern speed and the material speed. Such a scenario is true for bars. Models of the interstellar medium in spiral galaxies suggest that the presence of a strong central bar generates an inflow of gas, which accumulates at an inner resonance (known as the Inner Lindblad resonance—see chapter 3 for a definition). This is fuel for potential star formation and it is known that bars are associated with star formation in spiral galaxies. These models use hydrodynamic solutions to determine the gas flow through bars and find that strong shocks form on the leading edges of strong bars. This has been seen in real galaxies, where dust lanes, conventionally used as tracers of shocks, are seen on the leading edges of bars, when it is assumed that spiral structure is trailing. The strength of these shocks and the amount of gas flowing through them is thought to lead to a burst of star formation in the centers of barred galaxies. The modeled gas flow shows that material can originate in the disk and be forced into the bar, where it encounters the shock and then travels along the shock to the galaxy center. The results of these simulations have been used to develop the ideas of secular evolution of disk galaxies, where this flow of material is used as a mechanism for internal evolution of galaxy structures.

2.9 Secular evolution

Secular evolution is essentially a mechanism for transfer of material from the disk to the central regions of galaxies (i.e., the bulge or nucleus) via angular momentum transfer to the outer regions and redistribution of gas. We have already seen that spiral galaxy disks are well described by exponential light profiles. N-body simulations of angular momentum transfer also yield exponential light profiles. However, they also produce an exponential light profile in the central regions (i.e., for bulges) due to non-axisymmetric disturbances, which induce an inward radial flow of material.

Bars improve the efficiency of transport of material to the central regions, and bars are easily formed through naturally occurring instabilities in a cold self-gravitating disk of stars. Bars enhance secular evolution by funneling fuel from the disk to the central regions. As a result, disk material will be heated and thus its scale height is increased to 1–2 kpc above the plane of the disk via resonant scattering of stellar orbits by the bar instability. A bulge-like component with a nearly exponential light profile will emerge due to relaxation induced by the bar. The properties of this bulge are directly coupled to the relative timescales for star formation and angular momentum transfer. This model is expected to produce correlated scale lengths and colors between the disk and bulge. Observational evidence in favor of this model can be seen in the form of deviations from the $r^{1/4}$ law for bulge light profiles (especially when they are found to be better fit by exponential laws), color gradients and line strength gradients in the disk, and correlations between the bulge stellar populations and the disk stellar populations. Nuclear rings of star formation are also often seen in barred spiral galaxies. This is interpreted as gas accumulation near or at the inner Lindblad resonance as the bar 'funnels' material inwards.

Gas redistribution by the bar can cause its own destruction. Secular accumulation of only 1–3 percent of the total stellar disk mass near the center of the galaxy is sufficient to induce dissolution of the bar to a lens or triaxial component and later into a spheroid (i.e., bulge). At least two-thirds of spiral galaxies have bars, as noted from optical and infrared imaging. Most have probably harbored a bar at some point in their evolution. After bar formation, thickening, and eventual dissolution, accretion of central dissipationless material can only occur via stellar accretion, starburst activity, or resonant scattering caused by spiral arm resonances (which is a minor effect). Secular evolution is a mechanism for producing small, central accumulations of material, particularly in later-type spiral galaxies. Larger bulges cannot form in this way, since the energy required to heat up the central regions to form a large bulge in earlier-type spiral galaxies is greater than the total mechanical energy available in the disk. Secular evolution can still be used to accumulate more material in the central regions, and potentially grow a larger bulge; it just is possible to initially create a large bulge via this mechanism.

2.10 Spiral structure and the interplay between structures in disk galaxies

Regular spiral structure can be seen in galaxies out to intermediate redshifts, i.e., $z \approx 1.0$ and maybe slightly further, when the Universe was less than half its current age. This suggests that spiral structure is a long-lived phenomenon. If that is the case, then any model of spiral structure must be able to maintain a long-lived pattern, or at least a recurring transient pattern. This is the driving motivation behind the models and theories described in the coming chapters in this book.

2.11 The rest of this book

The following chapters in this book look at different types of models that describe spiral structure. A set of models described as density wave models will be described in chapter 3. In chapter 4, the focus is on other theories of spiral structure that do not fit into the typical density wave classification. Chapter 5 focuses on star formation in spiral galaxies and how it can be affected by spiral structure models and chapter 6 focuses on how spiral structure can be used to indirectly infer other information about spiral galaxies. Finally, chapter 7 provides a summary.

Suggested further reading

Begeman K G *et al* 1991 *Mon. Not. R. Astron. Soc.* **249** 523–37
Buta R and Combes F 1996 *Fundam. Cosm. Phys.* **17** 95–281
de Jong R S 1996 *Astron. Astrophys.* **313** 45–64
de Vaucouleurs G 1953 *Mon. Not. R. Astron. Soc.* **133** 134–61
Graham A W and Driver S P 2005 *Publ. Astron. Soc. Aust.* **22** 118–27
Hubble E P 1963 *Realm of the Nebulae* (New Haven, CT: Yale University Press)
Freeman K 1966a *Mon. Not. R. Astron. Soc.* **133** 47–62
Freeman K 1966b *Mon. Not. R. Astron. Soc.* **134** 1–14
Freeman K 1966c *Mon. Not. R. Astron. Soc.* **134** 15–23
Seigar M S and James P A 1998a *Mon. Not. R. Astron. Soc.* **299** 672–84
Seigar M S and James P A 1998b *Mon. Not. R. Astron. Soc.* **299** 685–97

Chapter 3

Density wave theories of spiral structure

Spiral Structure has proven to be one of the more obstinate problems in astronomy and astrophysics. Swedish astronomer, Bertil Lindblad (picture in figure 3.1), struggled with this from 1927 until his death in 1965. He correctly recognized that spiral structure arises through the interaction between the orbits and the gravitational forces of stars in the disk, and he thus investigated the problem using stellar dynamics.

Shortly before Lindblad's death, C C Lin and Frank Shu recognized that spiral structure in a galaxy disk could be regarded as a density wave, a wavelike oscillation that propagates through the disk in a similar way to waves propagating through guitar strings. This became known as Spiral Density Wave Theory.

In this chapter, the winding problem is first introduced. This explains that if spiral arms are material arms then they wind up in a very short amount of time. There then follows a discussion of various models that have tried to solve this problem, by explaining a mechanism that describes spiral structure that will not wind up. This then leads to a discussion on how this affects star formation in the disks of galaxies.

3.1 The winding problem

Imagine an arm in a spiral galaxy that points radially outwards like a spoke on a bicycle at some initial time $t = 0$ so that $\phi = \phi_0$ (where ϕ is the azimuthal angle). The disk rotates differentially with angular velocity $\Omega(r)$, where r is the radial distance from the galaxy center. Therefore, the arm will not remain radial as the disk rotates. The equation of the arm is given by equation (3.1).

$$\phi(r,\ t) = \phi_0 + \Omega(r)t \tag{3.1}$$

The pitch angle is defined as the angle between a tangent to the spiral arm and a tangent to a circle with a center that is coincident with the galaxy center. For a

doi:10.1088/978-1-6817-4609-8ch3

Figure 3.1. Bertil Lindblad (1895–1965) was a Swedish astronomy, graduate of the University of Uppsala in Sweden, and Director of the Stockholm Observatory from 1927–65. Image credit: Stockholm Observatory, courtesyof Dr Per Olof Lindblad.

logarithmic spiral, the pitch angle remains constant with radius. The pitch angle, i, is given by

$$\cot i = r \left| \frac{d\phi}{dr} \right| \tag{3.2}$$

which can be rewritten as

$$\cot i = rt \left| \frac{d\Omega}{dr} \right| \tag{3.3}$$

For tightly wound arms, the pitch angle i is small. This angle can then be related to the separation between arms, Δr. If at some azimuthal angle, ϕ, the arms are located at r and $r + \Delta r$ then,

$$2\pi = |\Omega(r + \Delta r) - \Omega(r)|t \tag{3.4}$$

So, when Δr is very small (i.e., when $\Delta r \ll r$), we get

$$\Delta r = \frac{2\pi r}{\cot i} \tag{3.5}$$

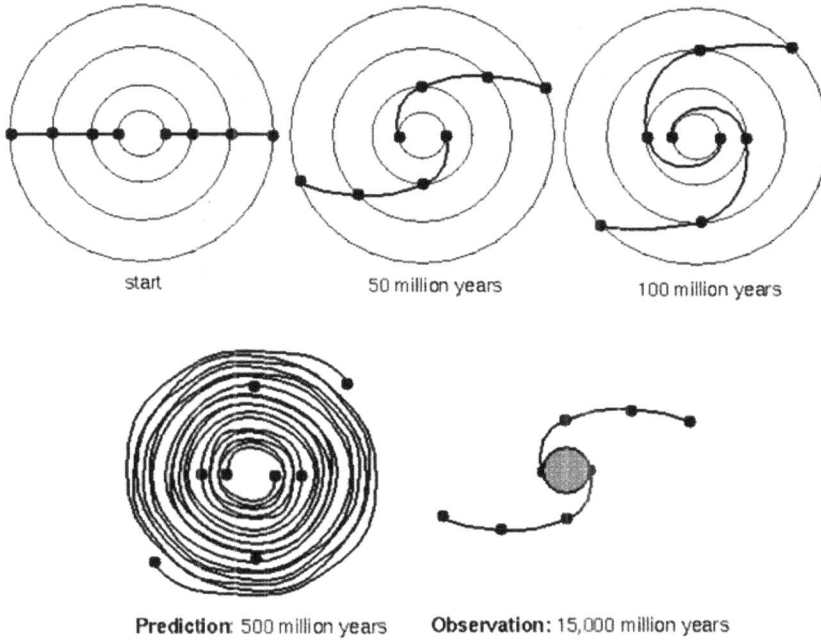

Figure 3.2. The winding problem. If spiral structure is the result of differential rotation in disk galaxies, then the spiral pattern is destroyed within 500 million years.

For a typical galaxy with a flat rotation curve, $\Omega r = v_c = 220 \text{ km s}^{-1}$, $r = 10 \text{ kpc}$, and $t = 10^{10}$ years. Under these circumstances, the pitch angle is $i = 0.25°$ and the separation between the arms is, therefore, $\Delta r = 0.28 \text{ kpc}$. This implies that the spiral pattern is much too tightly wound to be observable. In other words, if material making up the spiral arms remains in the arm, differential rotation of the disk winds up the arm in a small time compared to the age of the Galaxy. This is known as the 'winding problem' and was first noted by Elwira Wilczynski in 1896, and it is described diagrammatically in figure 3.2. A number of solutions have been proposed to overcome this problem, and this will be the focus of this chapter.

3.2 Spiral density wave theory

3.2.1 Lindblad's kinematic spiral waves

Lindblad (1961, 1963) realized that spiral structure arises due to the interaction between stellar orbits and the gravitational fields of galaxies and investigated this problem using stellar dynamics.

The radius of a stellar orbit in the disk of a galaxy is a periodic function of time with period T_r. During this period, the azimuthal angle increases by $\Delta\phi$. These quantities are related to the radial and azimuthal frequencies, ω_r and ω_a by $\omega_r = 2\pi/T_r$ and $\omega_a = \Delta\phi/T_r$.

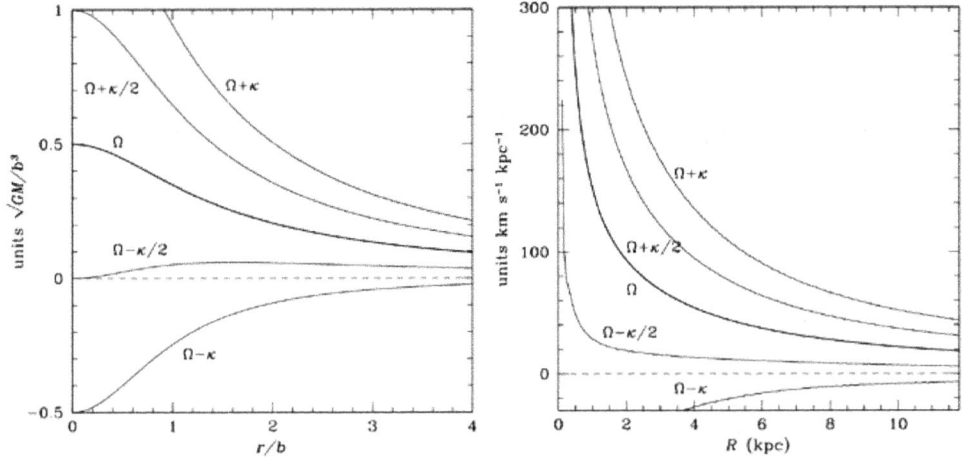

Figure 3.3. Behavior of $\Omega - n\kappa/m$ for the Bachall and Soneira (1980) model for the Milky Way Galaxy (left panel) and the isochrones potential (right panel). Reproduced from (Binney and Tremaine (2008) with permission of Princeton University Press.

In a frame rotating at angular speed Ω_p (the pattern speed), in one radial period the azimuthal angle increases by $\Delta\phi_p = \Delta\phi - \Omega_p T_r$. The pattern speed, Ω_p, can therefore be chosen so that the orbit is closed. If $\Delta\phi_p = 2\pi n/m$, where n and m are integers, the orbit closes after m radial oscillations. Here,

$$\Omega_p = \omega_a - \frac{n\omega_r}{m} \approx \Omega - \frac{n\kappa}{m} \tag{3.6}$$

where ω_a and ω_r have been approximated by their values for nearly circular orbits, Ω is the circular frequency and κ is the epicyclic frequency.

Figure 3.3 shows the behavior of $\Omega - n\kappa/m$ in the epicyclic approximation for several values of m and n. The figures show two representative rotation curves, the Bachall and Soneira (1980) model for the Milky Way Galaxy and the isochrone potential.

Lindblad noted that while most of the $\Omega - n\kappa/m$ curves varied rapidly with radius, the $n = 1$, $m = 2$ (or $n = 2$, $m = 4$ etc) curve is relatively constant over the galaxy disk. Suppose that the pattern speed $\Omega_p = \Omega - \kappa/2$. In a frame rotating at Ω_p, orbits are exactly closed and so a nested aligned set of orbits can be set up (see figure 3.4a). Stars in these orbits will create a bar-like wave pattern. By rotating the orbits of the ellipses, leading or trailing spiral density waves are created (figures 3.4b and c). From an inertial frame, the pattern rotates at Ω_p, which is called the pattern speed.

In real galaxies, $\Omega - \kappa/2$ is not exactly constant and so orbits are not exactly closed. As a result, the pattern created by such density waves can still be subject to the winding problem although they wind up about 5 times slower than material waves. However, this does show a certain amount of resistance to the winding problem and it explains the dominance of two-armed spiral patterns in Grand Design spiral galaxies. Density waves such as these are termed kinematic density waves because they involve the kinematics of orbits in a central potential.

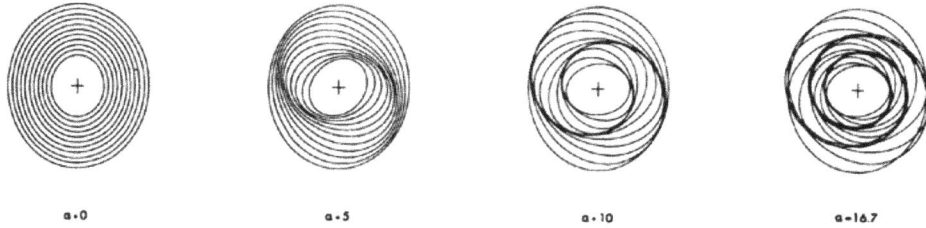

Figure 3.4. Arrangement of closed orbits in a galaxy with $\Omega - \kappa/2$ independent of radius to create bars and spiral patterns. Reproduced from Kalnajs (1973) with permission of Cambridge University Press.

3.2.2 Resonances

Stellar orbits have natural resonant frequencies. If the gravitational field generated by spiral structure perturbs an orbit at or near a resonant frequency, the response of the orbit will be large. At what frequencies do these resonances occur?

An arbitrary gravitational potential is written as $\Phi_1(R, \phi, t)$. The most important potentials are stationary in a rotating frame and are written as $\Phi_1(R, \phi, t) = \Phi(R, \phi - \Omega_p t)$, where Ω_p is the pattern speed of the potential. Systems that generate such potentials are bars, satellite galaxies and spiral patterns with well-defined pattern speeds.

What is the effect of such a potential on a disk of stars on near-circular orbits? The potential can be decomposed into a series of terms that are proportional to $\cos[m(\phi - \Omega_p t) + \text{constant}]$. Orbits in potentials of this form have resonances that occur when the circular frequency Ω and the epicyclic frequency κ satisfy the following conditions:

    (i) when $\Omega = \Omega_p$ we get a corotation resonance;
    (ii) when $\Omega = \Omega_p \pm \frac{\kappa}{m}$ we get Lindblad resonances.

Figure 3.3 shows that typical galaxies can have zero, one, or two Lindblad resonances. If a line of constant pattern speed is overlaid on figure 3.3a, the point at which this line intersects with the lines indicated by Ω and $\Omega \pm \kappa/m$ indicates where the resonances occur and in figure 3.3a this indicates 3 resonances.

When $n = \pm 1$, equation 3.6 for a stationary kinematic wave is identical to the condition for a Lindblad resonance. Near resonance, a weak perturbation with m-fold symmetry will produce a strong response with the same symmetry.

It is general believed that inner Lindblad resonances (ILRs) can lie within the central regions of barred galaxies (see e.g., Seigar 2002). The main arguments in favor of this proposal are:

1. Dust lanes are observed along the leading edges of bars. Models have shown that in bars with an ILR, the dust lane is offset further toward the leading edge, but without an ILR the dust lanes are central.
2. Nuclear rings of star formation have been observed (e.g., Seigar 2002). At the ILR, the perturbation leads to enhanced activity such as a burst of star formation.

3. Rings of molecular gas have been observed, which is another indication of enhancement at an ILR.

M 100 is a late-type barred spiral galaxy with a prominent ring of star formation within the vicinity of its ILR (Pierce 1986; Knapen *et al* 1995). M 74 has an enhancement of CO and a ring of star formation near its ILR (Seigar 2002) although it does not harbor an obvious bar.

3.2.3 The Lin–Shu hypothesis

In 1964, c.c. Lin and Frank Shu postulated that spiral structure is a wave pattern that remains quasi-stationary in the frame of reference rotating around the galaxy center at a constant angular velocity at all radii. Such a 'density wave' provides a spiral pattern, which underlies the observable concentrations of stars and gas. The pattern can therefore be maintained over the whole disk.

If there is some kind of instability in the disk (possibly gravitational), then this will take the form of motions within the disk. In their model, Lin and Shu (1964) used an infinitesimally thin disk so that the surface density is approximately the projected density in the gravitational plane. They also assumed that gravity is in balance with the circular and random velocities. They found that it is convenient to have a singularity of the density distribution at the center of the Galaxy, provided that the total mass is finite. The following is an outline of the analysis provided by Lin and Shu in their 1964 paper.

The basic equations of stellar dynamics in terms of the distribution function in phase space are shown in equations 3.7 through 3.10 (see Chandrasekhar 1960) as follows,

$$\mu_t + r^{-1}[(r\mu u)_r + (\mu v)_\theta] = 0 \tag{3.7}$$

$$u_t + uu_r + (v/r)u_\theta - v^2/r = \phi_r \tag{3.8}$$

$$v_t + uv_r + (v/r)v_\theta + uv/r = \phi_\theta/r \tag{3.9}$$

$$\phi_{rr} + \phi_r/r + \phi_{\theta\theta}/r^2 + \phi_{zz} = -4\pi G\mu(r, \theta)\delta(z) \tag{3.10}$$

where μ is the surface density and ϕ is the negative of the gravitational potential. Equations (3.7) to (3.10) are shown in cylindrical coordinates (r, θ, z).

In initial equilibrium, $\mu = \mu_0(r)$, $u = 0$, and $v = V(r) = r\Omega(r) > 0$. The disk can be disturbed such that $\mu = \mu_0(r) + \mu'(r, \theta, t)$. Consider a small disturbance so that μ' can be linearized. Using equations (3.7)–(3.10) this gives:

$$\mu_t' + \Omega\mu_\theta' + \frac{1}{r}(r\mu_0 u')_r + (\mu_0/r)v_\theta' = 0 \tag{3.11}$$

$$u_t' + \Omega u_\theta' - 2\Omega v' = \phi_r' \tag{3.12}$$

$$v'_t + \Omega v'_\theta + (\kappa^2/2\Omega)u' = \phi'_\theta/r \qquad (3.13)$$

$$\phi'_{rr} + \phi'_r/r + \phi'_{\theta\theta}/r^2 + \phi'_{zz} = -4\pi G\mu'\delta(z) \qquad (3.14)$$

where κ is the epicyclic frequency, such that $\kappa^2 = 4\Omega^2[1 + (r/2\Omega)(d\Omega/dr)]$.

This yields a solution of the type:

$$\mu' = Re\{\mu^{(1)}(r)\exp[i(\omega t - n\theta)]\}, \qquad \omega = \omega_r + i\omega_i \qquad (3.15)$$

where n is an integer and unstable modes are given by $\omega_i < 0$.

The solution in equation (3.15) is of the form of a density wave with as spiral form, such that,

$$\mu^{(1)}(r) = S(r)\exp[i\Phi(r)] \qquad (3.16)$$

where S and Φ are real functions. This therefore leads to the following:

$$\mu'(r, \theta, t) = S(r)\exp(-\omega_i t\cos[\omega_r t - n\theta + \Phi(r)]). \qquad (3.17)$$

As radial distance, r, increases $S(r)$ varies slowly and $\Phi(r)$ varies rapidly, and so equation (3.17) induces a spiral impression in the density distribution at any time. The form of this spiral is given by

$$\theta = \frac{1}{n}[\Phi(r) + \text{constant}] \qquad (3.18)$$

where n is the number of arms. If $\Phi(r) < 0$, the arms are trailing and if $\Phi(r) > 0$, the arms are leading.

If $\Phi(r)$ is rapidly varying, then the Poisson equation can be integrated using an asymptotic process, such that

$$\phi^{(1)}_{rr} + \phi^{(1)}_r/r - n^2\phi^{(1)}_{zz}/r^2 + \phi^{(1)}_{zz} = -4\pi G\mu^{(1)}(r)\delta(z) \qquad (3.19)$$

$$\mu^{(1)} = S(r)\exp[i\lambda f(r)] \qquad (3.20)$$

$$\phi^{(1)}(r, z) = \psi(r, z, \lambda)\exp[i\lambda h(r, z)] \qquad (3.21)$$

where λ is large and real, ψ is assumed to have an expansion in inverse powers of λ, and all of these functions may be complex.

An asymptotic solution is needed for the Laplace equation for $z > 0$ and $z < 0$, bounded at infinity, continuous at $z = 0$ and fulfilling $[\phi_z] = -4\pi G\mu^1(r)$ at $z = 0$, that can be derived from equation (3.19). To a first approximation,

$$\phi^{(1)}_r = 2\pi iG\varepsilon\mu^{(1)}, \qquad \phi^{(1)}_\theta = O(\mu^{(1)}/\lambda) \qquad (3.22)$$

where, according to the sign of the real part of $f'(r)$, $\varepsilon = \pm 1$. We must therefore determine the phase factor using equation (3.23).

$$\lambda f'(r)\varepsilon = \frac{[\kappa^2 - (\omega - n\Omega)^2]}{2\pi G\mu_0} \tag{3.23}$$

Therefore, a solution of the type considered in equation (3.15) is only possible if the following inequality is true:

$$\kappa^2 + \omega_i^2 - (\omega_r - n\Omega)^2 > 0. \tag{3.24}$$

Density waves are propagated primarily by gravitational forces, but should be modified by differential rotation, when non-linear terms omitted from equations (2.11) to (2.14) are included. This is analogous to fluid motion distorting acoustic waves. A density increase in the direction of the wave propagation tends to be accentuated into a compression shock, but a decrease would be smoothed out by the motion of the fluid. Therefore, only trailing waves are stable in the presence of non-linear effects. Trailing arms are those that point away from the direction of rotation.

Non-axisymmetric disturbances can propagate around the disk without change of shape even in the presence of differential rotation. For $r_1 < r < r_2$ (i.e., between the inner and outer Lindblad resonances) in which equation (3.24) applies, the geometrical form of the spiral pattern (from equations (3.18) and (3.23)) is given by,

$$n(\theta - \theta_0) = \int_{r_0}^{r} \frac{[\kappa^2 + \omega_i^2 + (\omega_r - n\Omega)^2]}{(2\pi G\mu_0)} dr. \tag{3.25}$$

If there is comparatively greater concentration of mass in the center, the density of the disk, μ_0, is relatively smaller. Equation (3.25) predicts tighter spirals for such cases, and this is observed in Sa-type galaxies. A more even distribution of matter implies loosely wound arms as seen in Sc galaxies. Indeed, observations have shown that a weak correlation exists between Hubble type and pitch angle (Kennicutt 1981; Seigar and James 1998). And this supports the prediction of equation (3.25).

The Lin–Shu hypothesis basically states that the total stellar population, which has various degrees of velocity dispersion, forms a quasi-stationary spiral structure. This is primarily due to the effect of gravitational instability as limited by velocity dispersion. Some important results follow from this:

1. Due to the spiral gravitational field, gas and young stars should form similar patterns on the scale of the radius of the disk.
2. Galaxies devoid of gas do not show spiral patterns.
3. Lindblad emphasized the constancy of $\Omega - \kappa/2$ over the disk. Lin and Shu got local resonances only at two radii where $\kappa^2 - (\omega - n\Omega)^2 = 0$ in the case of neutral waves. This implies that the traveling spiral gravitational field is in step with the local epicyclic motion.

In 1969, c.c. Lin and Frank Shu reported a development on their original model. This involved the deviation of the dispersion relation for density waves, the idea being that any wave obeys an equation, which relates its wavelength to its frequency, i.e.,

$$\frac{k_*}{|k|}(i - v^2) = F_v(x) + \frac{\mu_0}{\mu_*}F_v^{(g)}(x_g) \tag{3.26}$$

where k is the wavelength of the gas, k_* is the wavelength of the stars and is given by $k_* = \kappa^2/2\pi G\mu$, ν is the frequency and is constrained by $1 - \nu^2 > 0$ and F_ν and $F_\nu^{(g)}$ are reduction factors given by:

$$F_\nu^{(g)}(x_g) = \frac{1}{1 + x_g/(1 - \nu^2)} \tag{3.27}$$

$$F_\nu(x) = \frac{1}{1 + x/(1 - \nu^2)} \tag{3.28}$$

where $x_g = k^2 a^2/\kappa^2$ and $x = k_*^2 a_*^2/\kappa^2$ and a is the equivalent acoustic velocity.

Waves described by equation (3.26) have the property that they extend over a range of the galactic disk for which the conditions,

$$\Omega - \frac{\kappa}{n} < \Omega_p < \Omega + \frac{\kappa}{n} \tag{3.29}$$

are satisfied, where n is the number of arms, κ is the epicyclic frequency, Ω is the angular velocity of the stars and Ω_p is the pattern speed. This region is known as the *principal part of the spiral pattern*.

Figure 3.5 shows a graph of $\Omega(r)$, $\kappa(r)$, and $\Omega \pm \kappa/2$ for $n = 2$ according to the Schmidt model for the Milky Way Galaxy. The spiral pattern should extend from $r = 4$ kpc to beyond 20 kpc is $\Omega_p = 11 \text{ km s}^{-1} \text{ kpc}^{-1}$. For $n > 2$ this part would be quite limited in extent, whatever the value of Ω_p. So a two-armed spiral is predicted to be the strongest form of axisymmetric perturbation, and indeed, most Grand Design spiral galaxies are observed to have two arms. This argument does not hold for the outer parts of galaxies where it is possible for more arms to form (e.g., M 101).

3.2.4 Local stability of differentially rotating disks

In 1929, James Jeans took up the question of self-gravitating gas and found that under certain conditions it could be unstable enough to collapse under its

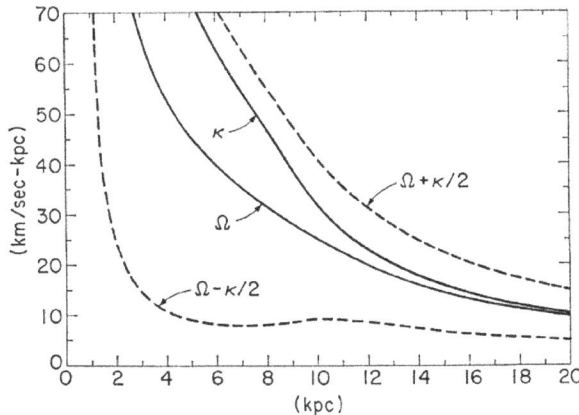

Figure 3.5. Rotation curves etc of the Milky Way Galaxy according to Schmidt.

own self-gravity. In order to show this, he considered an adiabatic gas with $p \propto \rho^{\gamma}$. A similar calculation can be applied to stellar systems.

The idea of instabilities is essentially a condensation of material, so in order for material to condense it is necessary to find out if gravity will cause collapse before velocity dispersion causes expansion. A characteristic time can be calculated for each process and compared to see which process is dominant. It turns out that on small scales, velocity dispersion is dominant, and on large scales, gravity is dominant. Jeans found a critical length above which gravitational instability becomes dominant. This is known as the Jeans length and is formulated as follows:

$$L_J = \sqrt{\frac{3\pi}{32} \frac{\langle v^2 \rangle}{G\rho}} \qquad (3.30)$$

If the size of a cluster of stars, $L > L_J$, then stars cannot avoid gravitational collapse (i.e., in a stellar medium characterized by given values of $\langle v^2 \rangle$ and ρ, all lengths greater than L_J are gravitationally unstable).

The situation in the disks of galaxies is different from the above problem due to the flatness of the system (instead of a spheroid assumed in the Jeans analysis) and more importantly, differential rotation. Velocities due to differential rotation are approximately proportional to ΔR and might prevent the collapse from taking place even at distances $>L_J$. Differential rotation therefore inhibits gravitational collapse on large scales. The question is: what happens in between? In 1964, Alar Toomre attempted to answer this question. Firstly, the Jeans length must be modified as follows to account for the difference between a disk and a sphere:

$$L_J = \frac{\pi}{8} \frac{\langle v^2 \rangle}{G\mu} \qquad (3.31)$$

where μ is the surface density of the disk. Toomre (1964) then investigated the balance between differential rotation and self-gravitation. Differential rotation manifests itself physically from the fact that a contracting region conserves angular momentum. This spins up and causes a centrifugal force that might inhibit further collapse. First, note that the average angular velocity with respect to a fixed system is B, the second Oort constant. Consider a region of original radius L_0. Its angular momentum at the edge, per unit mass, is $L_0^2 B$. If it contracts to radius L, conservation of angular momentum requires that $L^2 \Omega = L_0^2 B$, so that its angular velocity will be $\Omega = L_0^2 B / L^2$. Associated with this is a centrifugal acceleration, directed outwards, whose size is $a_r = L\Omega^2 = L_0^4 B^2 / L^3$. There is also a gravitational acceleration, directed inwards, whose size is $a_g = G\pi L_0^2 \mu / L^2$. If these are initially in balance, and a small perturbation occurs, is the system unstable to gravitational collapse?

Suppose the two are in balance at L_0. Impose a small contraction, $-dL$, such that,

$$da_r = \frac{3L_0^4 B^2}{L^4} dL \qquad (3.32)$$

$$da_g = \frac{2G\pi L_0^2 \mu}{L^3} dL. \tag{3.33}$$

The critical condition is that equations (3.32) and (3.33) are equal at $L = L_0$ and from this,

$$L_{\rm rot} = \frac{2G\pi\mu}{3B^2} \tag{3.34}$$

Thus the differential equation will stabilize against gravitational collapse for length scales $L > L_{\rm rot}$. A disk is therefore gravitationally unstable in the range of lengths $L_J < L < L_{\rm rot}$. The minimum condition for stability of a disk is that $L_{\rm rot} = L_J$ or,

$$\langle v^2 \rangle^{1/2} = \frac{4}{\sqrt{3}} \frac{G\mu}{B} \tag{3.35}$$

If the stars of a disk have a lower velocity dispersion than this, the disk will be unstable.

In the more elaborate and complicated analysis that Toomre undertook in 1964, this velocity dispersion, $\sigma_{\mu,\,\rm min}$, was derived as,

$$\sigma_{\mu,\,\rm min} = 3.36 \frac{G\mu}{\kappa} \tag{3.36}$$

where κ is the epicyclic frequency. The stability of disks is usually quoted as the ratio of the actual velocity dispersion, σ_μ, to that in equation (3.36), i.e.,

$$Q = \frac{\sigma_\mu}{\sigma_{\mu,\,\rm min}} = \frac{\sigma_\mu \kappa}{3.36 G\mu}. \tag{3.37}$$

Thus if $Q > 1$ then the velocity dispersion is high enough to prevent gravitational collapse, and if $Q < 1$ then gravitational collapse occurs. This condition is known as the *Toomre stability criterion*.

The Toomre stability criterion applies to only axisymmetric instabilities. The critical length, where $L_{\rm rot} = L_J$ is many kiloparsecs. Therefore, the violation of the Toomre stability criterion is not the only condition needed for the formation of spiral arms.

3.2.5 Swing amplification

Density wave theories predict that spiral patterns are trailing in nature and that any leading disturbance in a disk must eventually unwind.

This section explains the idea of the swing amplifier as proposed by Toomre (1981). This is a mechanism by which leading patterns unwind into trailing spirals.

Consider a stellar disk with constant Q and assume that a fraction $f/(1+f)$ of the radial force on the disk arises from a fixed halo component. The disk surface density is $\mu(r) = \mu_0 r_0 / r$ and the angular velocity is related as follows:

$$\Omega^2 = \frac{2\pi G\mu_0 r_0}{r^2}(1+f) \tag{3.38}$$

Linear perturbation theory can be used to follow the evolution of a leading wave with $Q = 1.5$ and $f = 1$. This is shown in figure 3.6. Within four rotation periods the wave unwinds and then turns into a trailing pattern and becomes more tightly wound. The amplitude of the trailing wave in frame 9 of figure 3.6 is about 20 times that of the leading wave in frame 1 of figure 3.6. The intermediate stages have even stronger trailing patterns. This is the result of *swing amplification*.

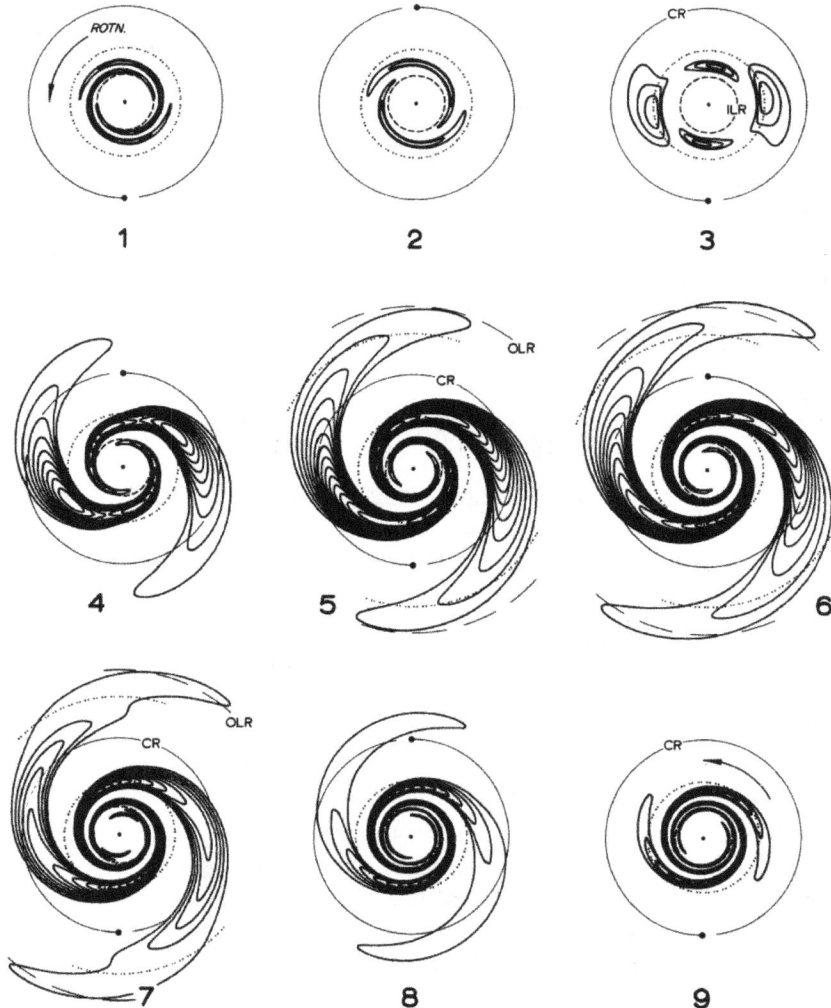

Figure 3.6. Evolution of a packet of leading waves in a stellar disk with $Q = 1.5$ and $f = 1$. Contours represent fixed fractional excess surface densities; the amplitude normalization is arbitrary. The time intervals between diagrams is one-half of a rotation period at corotation. Reproduced from Toomre (1981) with permission of Cambridge University Press.

3.2.6 Density wave theory and the classification of galaxies

Over the years there have been a lot of studies on the links between density wave theory and the classification of spiral galaxies. One of the earliest studies was by Roberts, Roberts and Shu (1975), who quantified the difference between structures with narrow, long, well-developed spiral arms, and those with broad 'massive', more fuzzy and patchy spirals. Their main result was that spiral galaxy classification depends only upon two parameters:

1. The total mass of the Galaxy divided by a characteristic dimension.
2. The degree of concentration of mass towards the galactic center (which in turn determines the pitch angle, i).

Together, these parameters determine ω_0, the velocity component of basic rotation normal to the spiral arms and this strongly influences the strength of the galactic shock wave.

It is the shock wave in the interstellar gas, which provides a triggering mechanism for the gravitational collapse of gas clouds into stars (a mechanism known as the large-scale shock scenario as investigated initially by W W Roberts in 1969). It can be shown that spiral structure with narrow 'filamentary' arms and broad 'massive' arms is dependent on,

$$\omega_0 = r(\Omega - \Omega_p)\sin i \qquad (3.39)$$

where i is the pitch angle and ω_0 is the angular speed with respect to the gas.

Therefore, if $\omega_0 > a$ where a is the effective acoustic speed of the interstellar gas, then most of the gas is supersonic and this yields strong shocks, which in turn give rise to narrow regions of high gas compression. This would therefore result in narrow, well-developed, visible spiral arms. However, if $\omega_0 < a$ then most of the gas is subsonic and this yields weak shocks and so weak, broad regions of relatively low gas compression and therefore patchy and fuzzy, visible spiral arms are formed.

In their paper, Robert, Roberts and Shu observed 24 galaxies and evaluated ω_0 for each of them. They then plotted this against the luminosity classification of each galaxy, as defined by van den Bergh (1960). It was van den Bergh who showed that a correlation existed between the intrinsic luminosity of a spiral galaxy and the degree to which spiral structure is developed. Roberts, Roberts and Shu, therefore, expected to see a good correlation in this plot and this can indeed be seen in figure 3.7. As already explained, a high rate of star formation is also expected with strong shocks. This is consistent with van den Bergh's correlation between intrinsic luminosity and the degree to which spiral structure is developed, as a high star formation rate is reflected through a high surface brightness. This was also seen by Kennicutt (1982), who presented a correlation between arm width and luminosity with a slope that was consistent with broadening by galactic rotation.

The pitch angle, i, is governed by the fundamental parameter $r_{0.5M}/r_{CO}$ where $r_{0.5M}$ is the half-mass radius and r_{CO} is the corotation radius, and so their ratio is a measure of the degree of mass concentration towards the galactic center. If $r_{0.5M}/r_{CO}$

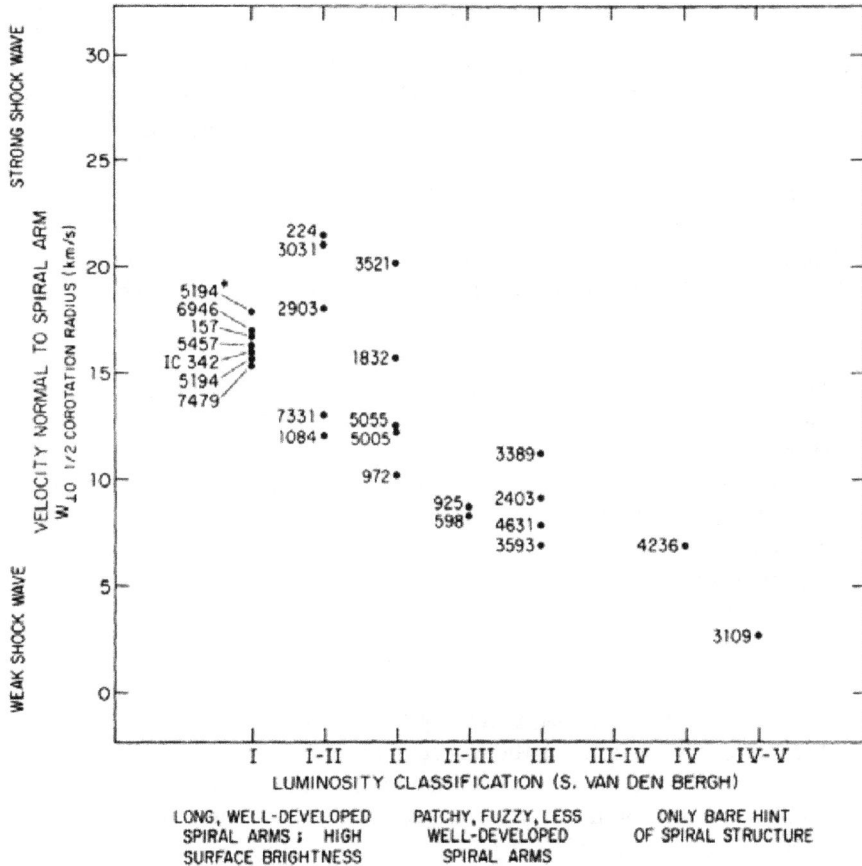

Figure 3.7. ω_0 and luminosity class. The trend for a sample of 24 galaxies from Roberts, Roberts and Shu (1975), indicative of the correlation between the velocity component of basic rotation normal to a spiral arm (ω_0) and shock strength on one hand, and luminosity classification and degree of development of spiral structure on the other. Those galaxies in which strong shocks are predicted are found to exhibit long, well-developed spiral arms; and those in which weak shocks are predicted are found to exhibit less-developed spiral structure. Copyright AAS. Reproduced with permission from Roberts, Roberts and Shu (1975).

is small then there is a major proportion of mass distributed with a high central concentration. This implies strong differential rotation and a small value for i and this yields a tightly wound spiral structure, i.e., earlier-type spirals such as Sa. If, on the other hand, $r_{0.5M}/r_{CO}$ is large, one would get a high value for i and therefore loosely wound spiral arms, as seen in Sc type galaxies. Figure 3.8 shows the correlation that Roberts, Roberts and Shu measured between pitch angle and Hubble type. The scatter in the correlation is probably because the self-gravity of the gas is not entirely negligible and so M_{gas}/M_{stars} may be important.

The predicted pitch angles of this model were tested by Kennicutt and Hodge (1982), who used Hα imaging of 17 of the galaxies from the Roberts, Roberts and Shu (1975) sample to measure their pitch angles. Kennicutt and Hodge found a good

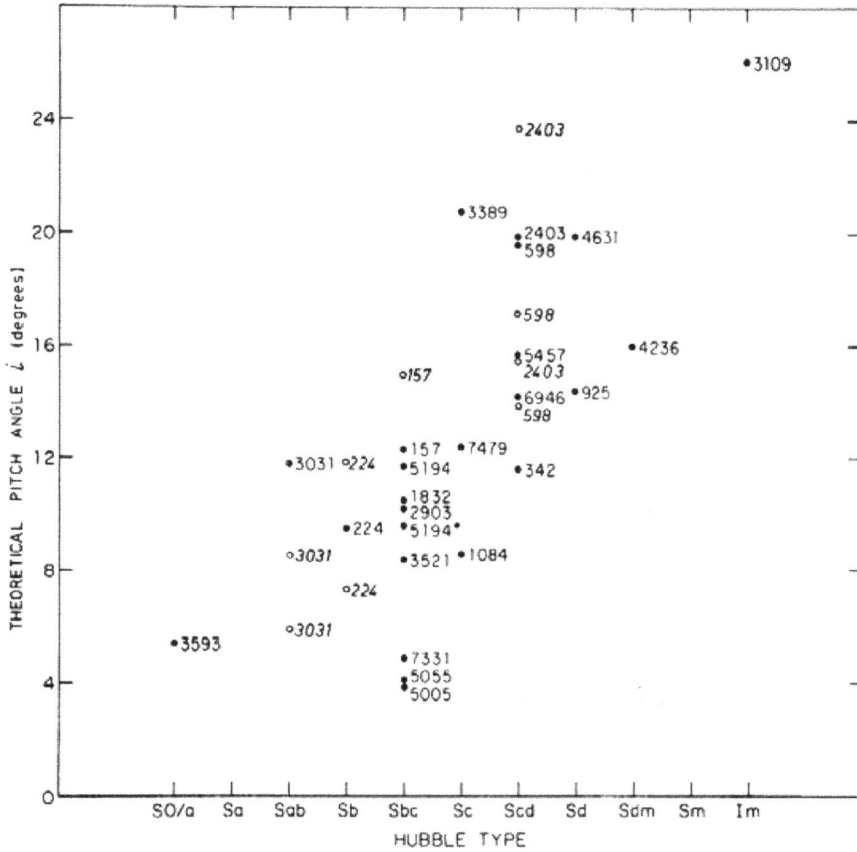

Figure 3.8. Theoretical pitch angle and Hubble type. The trend for a sample of spiral galaxies indicative of the correlation between theoretical pitch angle of the density wave pattern and Hubble type. Those galaxies whose models predict wave patterns with tightly wound arms are observed to be off relatively early type, S0/a—Sbc; and those galaxies whose models predict more open arms with large pitch angles are observed to be of relatively late type, Sc—Im. The dark dots indicate the cases adopted by Roberts, Roberts and Shu (1975). The open circles represent other possible choices of corotation radius. Copyright AAS. Reproduced with permission from Roberts, Roberts and Shu (1975).

correlation between their measured pitch angles and those predicted by the Roberts, Roberts and Shu (1975) model but they also found that the model systematically underestimated all of the pitch angles. Conversely, Seigar and James (1998) found no correlation between pitch angle and Hubble type or between pitch angle and light fraction in the disk (see figure 3.9).

A later study by Seigar *et al* (2005), found that a correlation exists between spiral arm pitch angle and rotation curve shear. Rotation curve shear depends on the slope of the rotation curve, with declining rotation curves having shear $S > 0.5$ and increasing rotation curves having shear $S < 0.5$. In essence, the shear is related to the central mass concentration, and it may be the cleanest mechanism for indirectly determining the mass concentration. The result by Seigar and collaborators has been

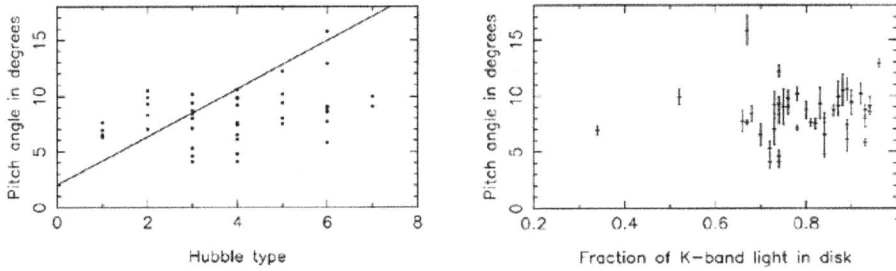

Figure 3.9. Left panel: pitch angle as measure in the near-infrared versus Hubble type, showing a lack of a correlation. The line represents the theoretical correlation from Roberts *et al* (1975). Right panel: pitch angle versus the fraction of near-infrared light in the disk. Figures from Seigar and James (1998).

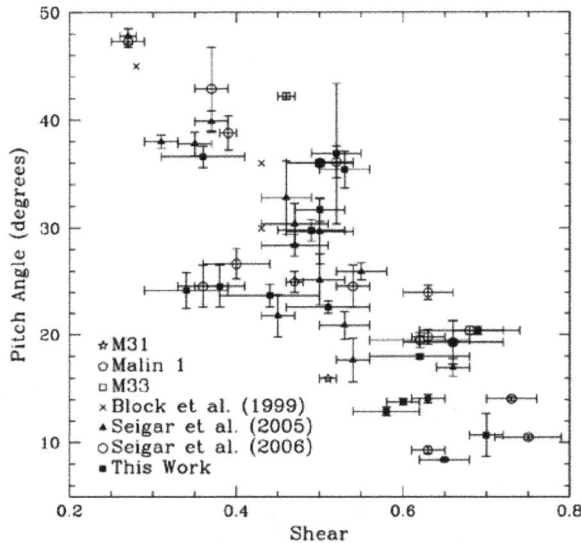

Figure 3.10. Spiral arm pitch angle versus rotation curve shear, showing a correlation. The filled triangles represent galaxies with data measured by Seigar *et al* (2005), the open circles are galaxies from Seigar *et al* (2006), the crosses are from Block *et al* (1999), the open square represents M33 (Seigar 2011), the open pentagon represents Malin 1 (Seigar 2008), the star represents M31 (Seigar *et al* 2008), and the filled squares represent the data from Seigar *et al* (2014).

confirmed in many follow-up investigations and figure 3.10 shows the latest version of the correlation.

3.2.7 The theory of global modes

The Modal theory, as established by Guiseppe Bertin and collaborators, is essentially an extension to density wave theory which describes large-scale spiral structure in Grand Design spiral galaxies as being long lasting and being associated with a few 'self-excited' global modes which trace intrinsic properties of the host galaxy. In this way, the disk of the galaxy can be treated as a resonant cavity with the boundaries at the inner Lindblad resonance (ILR) and the outer Lindblad resonance (OLR).

Global modes are composed of wavetrains propagating radially in opposite directions, thus creating a standing wave. Some kind of feedback is required from the center and this is provided by a refraction-like process, due to the bulge or a dynamically warmer inner disk. In the stellar disk, this feedback can be interrupted by an ILR, which can damp modes. In the gaseous disk, the related resonant absorption is only partial, but some feedback is guaranteed. Once this wavecycle is set up a global mode can be generated.

First-order modes ($m = 1$) are ILR free and may therefore be ubiquitous in sufficiently cool disks. Only some $m = 2$ modes will be ILR free (i.e., not damped). Higher-order modes should be damped by the ILR. Modal theories predict that near-infrared images (which highlight the old stellar population in the disks of galaxies) should reveal one and two-armed spiral structures and that the dynamics of the young-stellar disk imply that optical images should be dominated by higher-m modes and irregular, fast-evolving features. Indeed, several studies have shown that galaxies that are often classified as multi-armed or flocculent on the basis of their optical appearance, often are revealed to have a regular Grand Design spiral structure when imaged in the near-infrared, thus confirming this hypothesis.

In order for global modes to be long lasting, the cold dissipative interstellar medium (ISM) plays a self-regulatory role. In a stellar disk, there is conversion from ordered kinetic energy into random motions, which leads to heating of the disk. This would lead to eventual destruction of the spiral wave. However, in modal theory, the ISM can regulate the temperature of the disk by supplying a cooling mechanism.

Modal theory leads to a three-dimensional representation of the observed morphological categories (see figure 3.11). A spiral galaxy can be placed anywhere in the parameter space defined by three quantities:

1. The active disk mass
2. The gas content
3. The 'temperature' of the disk.

The transition from nonbarred (SA) to barred (SB) structure is dictated by the active disk mass. The transition along the Hubble sequence, from early (a) to late (c) is mostly dictated by the gas content and the transition from Grand Design (G) to flocculent (F) galaxies is determined by random motions in the disk. Grand Design structure requires small random motions, while a hot stellar disk leads to small-scale spiral structure that is gas supported and flocculent (see also the work by Elmegreen 1991 and Roberts 1992). In Grand Design galaxies, where the stellar disk is cool, the degree of coupling may only be partial. In hot stellar disks, full decoupling occurs and this describes flocculent spiral structure. In effect, the young stellar population disk and the old stellar population disk are decoupled and so, two entirely decoupled morphologies can coexist in the same galaxy. This has indeed been observed by several authors (Thornley 1996; Block *et al* 1994; Seigar and James 1998; Seigar *et al* 2003).

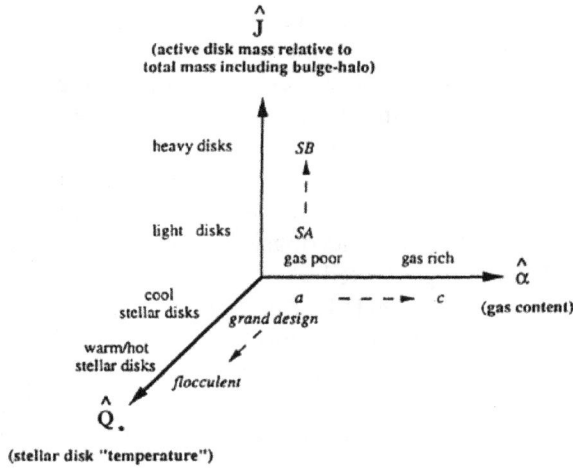

Figure 3.11. Framework for the classification of the morphologies of spiral galaxies on the basis of their intrinsic modal characteristics. Reproduced from Bertin (1991) with permission from Springer Nature.

3.2.7.1 Feedback and amplification of global modes

Consider a second-order differential equation of the form

$$\frac{d^2y}{dr^2} + g(r, \omega)y = 0 \tag{3.40}$$

with $-g(r, \omega)$ having the form sketched in figure 3.12.

The function g behaves like a potential well for a particle with zero energy. When appropriate boundary conditions are set at $r = 0$ and at $r \to \infty$, the solution yields only quantized values of ω.

The points where $g = 0$ are termed *turning points* and these exist at $r = r_{ce}$ and at $r = r_{co}$ (as shown in figure 3.12). These provide barriers through which the propagation of waves is inhibited. The situation is similar to that of quantum tunneling.

In order to set up a standing wave in an oscillatory system, it is necessary to have the participation of waves with opposite directions of propagation. Galaxy disks must therefore have a feedback mechanism, which is able to return an incoming wave as an outgoing wave. This should maintain the standing wave pattern.

In the model of equation (3.40) this feedback is present. The function is $g(r, \omega)$ at $r \approx r_{ce}$ represents a barrier. The turning point at r_{co} acts as a mirror for one of the two solutions allowed by equation (3.40).

3.2.7.2 Radiation boundary condition

The system defined by the function $g(r, \omega)$ sketched in figure 3.12 allows for oscillatory solutions at large radii. Using equation (3.40) in the context of the dynamics of galaxy disks, there is a situation where no energy sources are available from the outside, for an isolated galaxy. The appropriate condition is therefore that of an *outgoing wave*, i.e., the *radiation boundary condition*. This means that of the

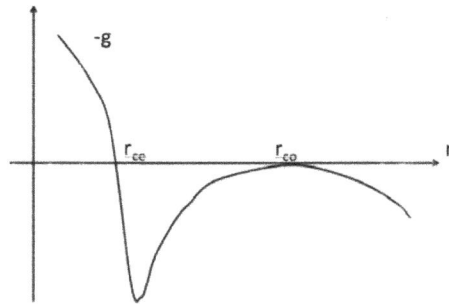

Figure 3.12. Illustration of the two turning point problem, indicating feedback, radiation boundary condition and overreflection.

two solutions available at large radii, the wave that carries energy outward should be chosen, and this prevents the galaxy disk from absorbing external energy.

This outgoing wave signal is expected to be absorbed either at the outer Lindblad resonance or by turbulent dissipation of the gas layer. Thus no signals from the outside are expected. This boundary condition can only be satisfied by trailing waves in a galaxy disk.

3.2.7.3 Overreflection

The radiation boundary condition leads to an important amplification mechanism called *overreflection*. Consider the double turning point at r_{co} as representative of the situation in a galaxy near corotation. A signal coming from the center of the Galaxy is therefore partly reflected at r_{co} and partially transmitted. In order to conserve energy and angular momentum across corotation, the reflected signal is stronger than the original wave. This amplification is shown in figure 3.13. The process is stronger where it is encouraged (middle panel in figure 3.14, where $g < 0$ at corotation). It is still present, but weak, when waves have to tunnel through r_{co} (right panel in figure 3.14, where $g > 0$ at corotation). The amplification takes place because waves are transferred from a region of negative action density to a region of positive action density.

3.2.7.4 Self-excitation of discrete global modes

The above description provides the justification for a spectrum of unstable, self-excited global spiral modes that are trailing. Only a few choices of ω will shape the g-curve so that the solution will satisfy both the radiation boundary condition and the regularity condition. The system behaves like a cavity for a laser process. A wave-signal is amplified in power as a result of overreflection at each cycle at a rate determined by the time it takes to travel from feedback at r_{ce} to the corotation radius. Therefore, the growth rate of a mode is inversely proportional to the group propagation time for a signal to move from r_{co} to r_{ce} and back again.

This holds unless there is something that can limit the feedback process. The inner Lindblad resonance acts in such a way and limits the number of stable modes present in a disk by damping modes of order 3 and higher.

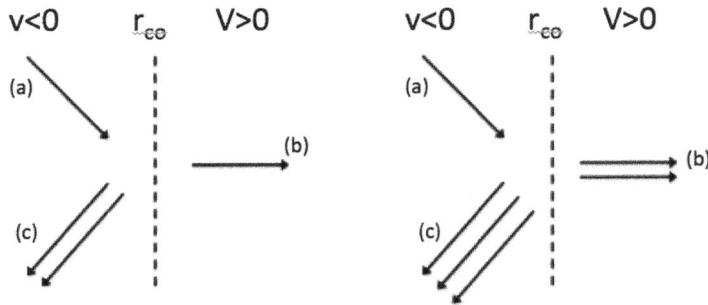

Figure 3.13. Process of overreflection illustrated for two cases of different levels of local stability at corotation. Left panel: case of marginal stability. Right panel: case of a system locally unstable at corotation.

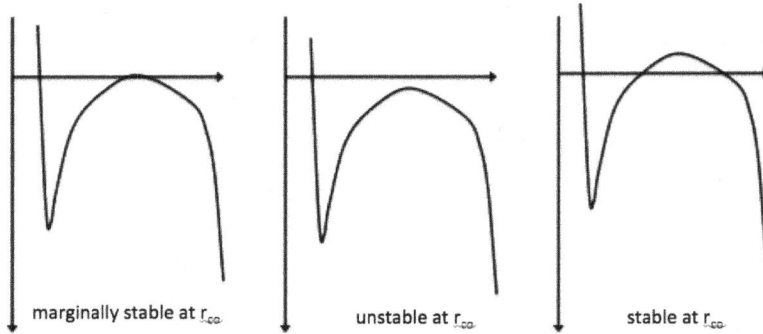

Figure 3.14. Function $(-g)$ can be seen like the potential for a particle with zero energy.

3.2.7.5 Observational evidence and the theory of global modes

The strongest evidence in favor of global modes comes from two-dimensional fast Fourier transforms of spiral structure in disk galaxies. These analyses often show that the spiral structure consists of multiple modes (see figure 3.15), and that, often the $m = 2$ mode dominates the structure. In some cases higher order even modes (e.g., $m = 4$) can dominate, but this tends to happen further out in the disks of galaxies. When an odd model dominates (e.g., $m = 1$ or $m = 3$) this is indicative of lopsidedness in a galaxy. Figure 3.16 shows an example of a galaxy that clearly has a three-armed structure and the Fourier analysis clearly shows that the $m = 3$ mode dominates. This particular case is inconsistent with modal theory.

3.2.8 Other density wave theories

3.2.8.1 Chaotic spiral arms

Theories of chaotic structure try to explain the more ragged structure that is seen in many spiral galaxies, i.e., those with short fragmented arms with no clear two-arm symmetry, e.g., NGC 2841 (see figure 3.17). This type of galaxy has been described as a 'swirling hotch-potch of spiral armss by Goldreich and Lynden-Bell (1965). Although they are less dramatic, they are more common than Grand Design spirals.

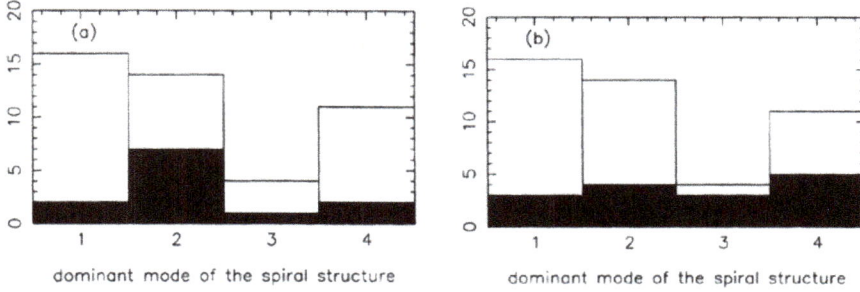

Figure 3.15. The occurrence of the strongest Fourier modes found in the disks of a sample of 45 galaxies studied by Seigar and James (1998), shown with (a) galaxies with nearby companions and (b) strongly barred galaxies.

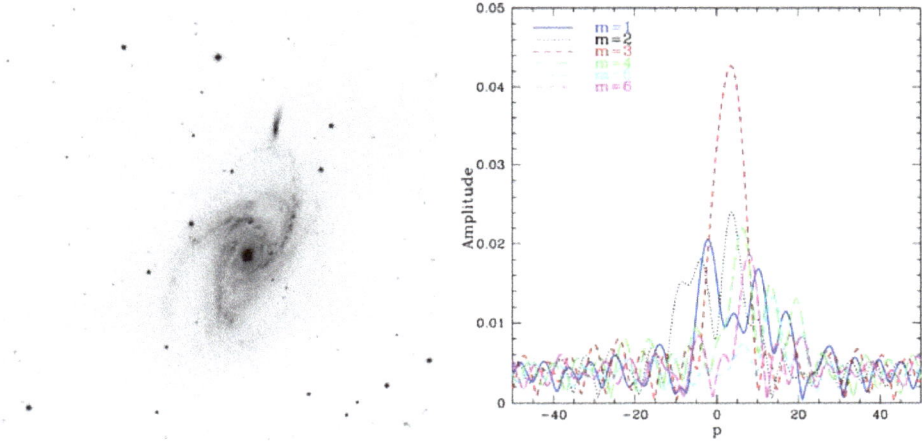

Figure 3.16. Left panel: B-band image of NGC 5054, clearly showing a three-armed structure. Right-panel: the amplitude of Fourier modes for a deprojected B-band image of NGC 5054. This shows that the three-armed spiral or $m = 3$ Fourier mode dominates the power spectrum as described by Davis *et al* (2012).

The chaotic theory for spiral structure claims that pieces of arms are constantly forming and dying. A local gravitational instability in the interstellar gas creates a patch of new stars, which are sheared by differential rotation into spiral form. These patches are sheared more and more and the brightest stars start to die and so they gradually fade, whilst new patches form elsewhere.

Assuming the Toomre stability parameter $Q = 1$, it is possible to calculate the length of a typical spiral patch in a galaxy. The most unstable wavelength for axisymmetric perturbations in a gas disk is:

$$\lambda_{\text{most-unstable}} = \frac{1}{2}\lambda_{\text{crit}} = \frac{2\pi^2 G\mu_g}{\kappa^2} \tag{3.41}$$

where μ_g is the local surface density for the gas and κ is the local epicyclic frequency.

Figure 3.17. The Sb galaxy NGC 2841. Image courtesy of NASA/JPL-Caltech.

This is similar for non-axisymmetric perturbations. A reasonable estimate for the length is given by:

$$L \approx \frac{1}{2} \lambda_{\text{most-unstable}} = \frac{\pi^2 G \mu_g}{\kappa^2} \tag{3.42}$$

For the Solar neighborhood, $\mu_g = 5$ M$_\odot$/pc^2 and $\kappa = 36$ km s^{-1} kpc^{-1}. Therefore, $L = 0.2$ kpc, which agrees reasonably with observations.

This implies that star formation is self-regulating. Cooling and infall of material reduce Q in the interstellar gas. This leads to a gravitational instability, which sparks star formation. The largest stars die in supernova explosions. This heats the interstellar gas, and as a result its velocity dispersion increases and Q increases. This inhibits the growth of instabilities that create new stars.

3.2.8.2 Tidal arms

The Kormendy–Norman hypothesis states that Grand Design spiral structure occurs mainly within the radius in which solid-body rotation occurs, thus over-coming the winding problem. They also claim that any galaxy exhibiting Grand Design structure outside this radius does so because of the effects of a tidal potential due to either a central bar or a nearby companion.

In an attempt to test this hypothesis, Seigar *et al* (2003) investigated near-infrared images of 17 nearby spiral galaxies. They found that in every galaxy, the spiral arms

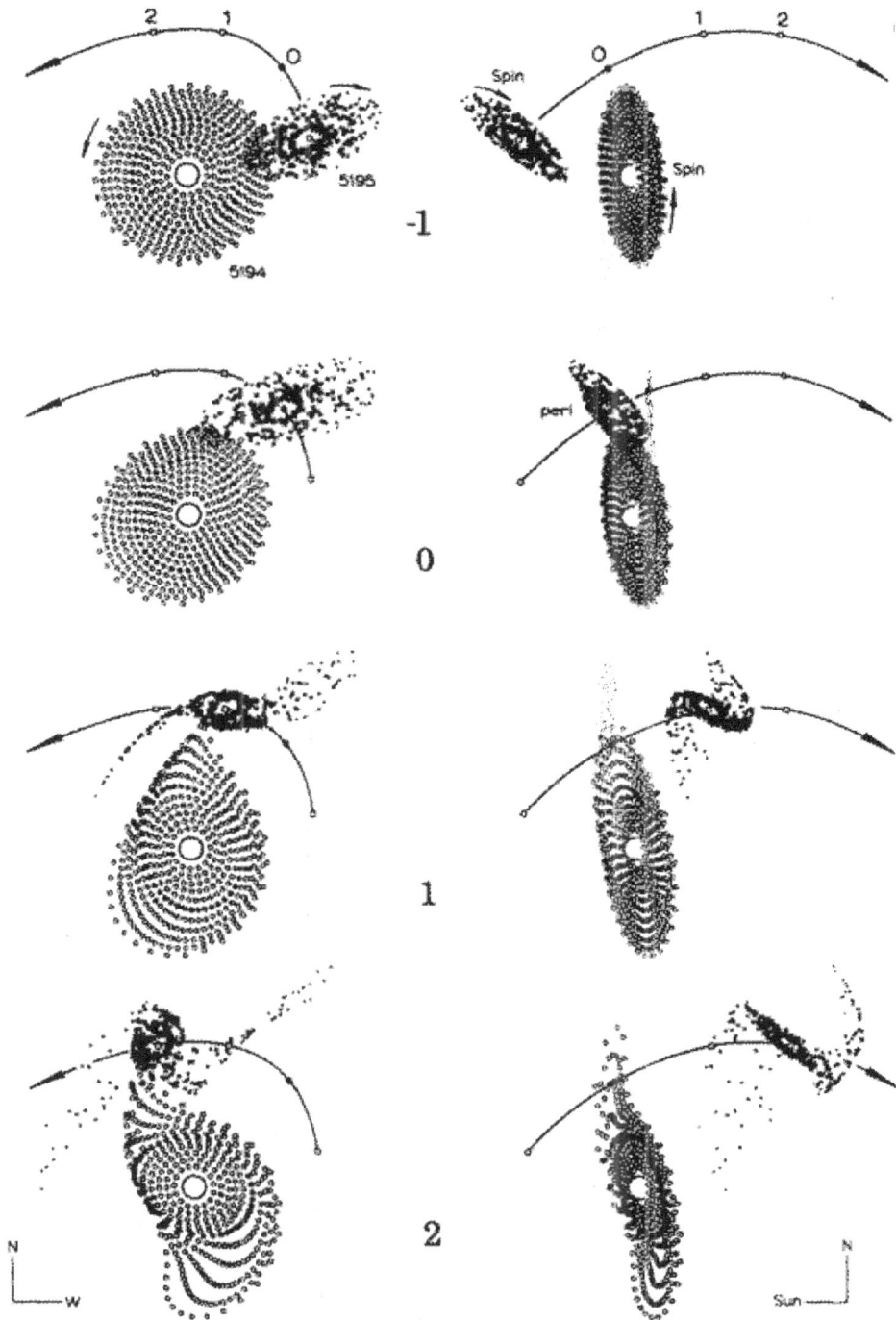

Figure 3.18. The evolution of spiral structure in the Messier 51/NGC 5195 system as modeled by Toomre and Toomre (1972). Reproduced from Toomre (1978) with permission from Springer Nature.

Figure 3.19. The Whirlpool Galaxy (spiral galaxy Messier 51 and its companion NGC 5195), classic spiral galaxies in the Canes Venaciti constellation. Image courtesy of NASA/ESA Hubble Space Teleascop.

extended beyond the turnover radius in their rotation curves. Only 11 of the galaxies showed strong evidence of a bar or nearby neighbor. From this study, it appears that galaxies can exhibit Grand Design spiral structure beyond the turnover radius.

One of the strongest cases for tidal spiral arms is Messier 51, which has a companion galaxy, NGC 5195. Toomre and Toomre (1972) investigated the possibility that the spiral structure in Messier 51 is the result of its interaction with NGC 5195. They modeled Messier 51 and its companion each as a disk of test particles surrounding a central point mass. Figure 3.18 shows that this model successfully produces the outer arms, but they cannot be traced into the center (compare this with the real image of Messier 51 shown in figure 3.19). This is probably due to the absence of self-gravity in the test particle disk.

The spiral is a density wave, but it is transitory rather than long-lived. Such models can explain the location and strength of the dust lanes, radio continuum and HI arms, and kinks in the HI velocity contours. However, encounters like this are not common enough to account for all Grand Design spiral galaxies.

Suggested further reading

Bachall J N and Soneira R M 1980 *Astrophys. J. Suppl.* **44** 73–110

Bertin G 1991 *Proc. of the 146th Symp. of the Int. Astronomical Union* (Dordrecht: Kluwer) p 93

Binney J and Tremaine S 2008 *Galactic Dynamics* 2nd edn (Princeton, NJ: Princeton University Press)

Block D L, Bertin G, Stockton A, Grosbol P, Moorwood A F M and Peletier R F 1994 *Astron. Astrophys.* **288** 365–82

Block D L, Puerari I, Frogel J A, Eskridge P B, Stockton A and Fuchs B 1999 *Astrophys. Space Sci.* **269** 5–29

Chandrasekhar S 1960 *Proc. Natl Acad. Sci. USA* **46** 253–7

Davis B L *et al* 2012 *Astrophys. J. Suppl.* **199** 33

Elmegreen B G 1991 *Astrophys. J. Col.* **378** 139–56

Goldreich P and Lynden-Bell D 1965 *Mon. Not. R. Astron. Soc.* **130** 97–124

Jeans J H 1929 *The Universe Around Us* (New York: Cambridge University Press)

Kalnajs A J 1973 *Proc. Astron. Soc. Aust.* **2** 174–7

Kennicutt R C Jr 1981 *Astron. J.* **86** 1847–58

Kennicutt R C Jr 1982 *Astron. J.* **87** 255–63

Kennicutt R C Jr and Hodge P 1982 *Astrophys. J.* **253** 101–7

Knapen J H, Beckman J E, Heller C H, Shlosman I and de Jong R S 1995 *Astrophys. J.* **454** 623–42

Lin C C and Shu F H 1964 *Astrophys. J.* **140** 646–55

Lin C C, Yuan C and Shu F H 1969 *Astrophys. J.* **156** 721–46

Lindblad B 1961 *Stockholms Observatoriums Annal.* **21** 8

Lindblad B 1963 *Stockholms Observatoriums Annal.* **22** 5

Pierce M J 1986 *Astron. J.* **92** 285–91

Roberts W W 1969 *Astrophys. J.* **158** 123–43

Roberts W W, Roberts M S and Shu F H 1975 *Astrophys. J.* **196** 381–405

Roberts W W 1992 *Ann. New York Acad. Sci.* **675** 93–107

Seigar M S and James P A 1998 *Mon. Not. R. Astron. Soc.* **299** 685–97

Seigar M S 2002 *Astron. Astrophys.* **393** 499–502

Seigar M S, Chorney N E and James P A 2003 *Mon. Not. R. Astron. Soc.* **342** 1–7

Seigar M S, Block D L, Puerari I, Chorney N E and James P A 2005 *Mon. Not. R. Astron. Soc.* **259** 1065–76

Seigar M S, Bullock J S, Barth A J and Ho L C 2006 *Astrophys. J.* **645** 1012–23

Seigar M S 2008 *Publ. Astron. Soc.* **120** 945–50

Seigar M S, Barth A J and Bullock J S 2008 *Mon. Not. R. Astron. Soc. Col.* **389** 1911–23

Seigar M S 2011 *ISRN Astron. Astrophys.* **2011** 725697

Seigar M S, Davis B L, Berrier J and Kennefick D 2014 *Astrophys. J.* **795** 90

Thornley M D 1996 *Astrophys. J. Lett.* **469** L45–8

Toomre A 1964 Astron. *J. Col.* **139** 1217–38

Toomre A and Toomre J 1972 *Astrophys. J.* **178** 623–66

Toomre A 1978 Interacting systems *The Large Scale Structure of the Universe* (Proceedings of the Symposium) (Dordrecht: D Reidel) p109–16

Toomre A 1981 *The Structure and Evolution of Normal Galaxies; Proceedings of the Advanced Study Institute* (New York: Cambridge University Press) pp 111–36

Van den Bergh S 1960 *Astrophys. J.* **131** 215–23

Wilczynski E 1896 *Astrophys. J.* **4** 97–100

Chapter 4

Other theories of spiral structure

Some theories of spiral structure do not fall under the category of density waves, at least not in the traditional sense. A recently proposed model, known as Manifold Theory, requires a bar to drive the spiral structure seen in disk galaxies. Quite often, these alternate theories can only be used to explain spiral structure in a subset of spiral galaxies, and even then, they often to not explain observed structures (see for example, the Pinwheel galaxy in figure 4.1, which obviously cannot be explained by Manifold Theory since it does not have a bar). Nevertheless, this chapter focuses on these alternate ideas, so that the reader can understand their shortcomings.

Figure 4.1. An example of a spiral galaxy, the Pinwheel Galaxy, also known as Messier 101. Image courtesy of European Space Agency/NASA.

4.1 Manifold theory

Any theory of galactic spiral structure based on orbital dynamics must provide a mechanism by which the phases of the orbits, i.e., the angular positions of the apsides (the points on an elliptical orbit that align with its major axis), become correlated, so as to reproduce self-consistently the observed spiral pattern. Models for normal spirals based on stable periodic orbits of the x_1 family, with apocenters aligned along the spiral, were proposed by Contopoulos and Grosbol (1986, 1988), and indeed this was the original goal of Bertil Lindblad in the 1950s. In such models, the main spiral ends near the 4:1 resonance (Contopoulos 1985; Patsis, Contopoulos and Grosbøl 1991; Patsis *et al* 1994; Patsis and Kaufmann 1999), although weak extension can be found out to the corotation region. Such models are probably not applicable at all if the spirals are supported mostly by chaotic orbits, as is probably the case when a bar is present (Kaufmann and Contopoulos 1996). The dominance of chaotic orbits near and beyond the corotation radius results in these orbits being a key feature (Kaufmann and Contopoulos 1996) in successful self-consistent models of spiral structure in barred-spiral galaxies (Sparke and Sellwood 1987).

A central question in the study of disk galaxies is whether the bar drives spiral structure (e.g., Goldreich and Tremaine 1978; Athanassoula 1980; Schwarz 1984, 1985; see also section 4.3), or if the spiral structure is a recurrent pattern characterized that naturally arises due to the gravitational dynamics of a disk of stars (see, e.g., Sellwood 2000, and references therein). A relevant key question is what mechanisms generate the spiral pattern and also whether these mechanisms work with regular or with chaotic orbits or both. So what models can be used to generate phase correlations of chaotic orbits, which are also able to support a regular spiral pattern?

The set of chaotic orbits near corotation fill randomly a connected chaotic domain of the phase space. The character of the orbits can then be naively perceived to oppose any persistent phase correlation among these orbits. On the other hand, it is well known in the theory of dynamical systems that the effective randomness exhibited by chaotic orbits is to a certain extent only apparent. In certain cases, invariant manifolds can be set up for the unstable periodic orbits in the chaotic domain. This is key in understanding the chaotic dynamics in the corotation region.

The unstable invariant manifolds of the short-period unstable periodic orbits provide a mechanism yielding phase correlations of chaotic orbits that support, precisely, a spiral density wave beyond the bar. The intersections of invariant manifolds with configuration space define patterns quite similar to the spiral pattern of the observed density field. Theoretically speaking, it can be demonstrated that the angular positions of the apocenters of chaotic orbits for successive orbits remain correlated for long times. This phase correlation causes the manifold to support the spiral pattern. Finally, another important property of invariant manifolds is their structural stability, i.e., the manifolds retain their phase-space geometric structure under small perturbations of a system.

This is the basis of the manifold theory of spiral structure in barred spiral galaxies. One of the key differences between manifold theory and other density wave theories is that density wave theories (such as modal theory and the Lin–Shu hypothesis) predict a slight offset in pitch angles for spirals observed in different wavebands, whereas manifold theory expects the same pitch angles because stars are trapped in the manifold. In chapter 5, an analysis of multiple waveband data is presented and an offset between near-infrared spiral arms and blue-light spiral arms can clearly be seen (see, for example, figure 5.2). This would suggest that manifold theory, in general, is not supported by observations.

4.2 Magnetohydrodynamic theories

Before 1964, some astronomers believed that spiral structure was caused by a complex interaction between interstellar gas and the galactic magnetic field. A calculation by Binney and Tremaine (1987) tests this idea.

To summarize, the energy density due to the magnetic field, B, is $B^2/(8\pi)$ in Gaussian units, or $B^2/(2\mu_0)$ in MKS units. The kinetic energy density in a perturbation that imparts a velocity, Δv, to a gas of density ρ is given as $\frac{1}{2}\rho\Delta v^2$. Velocity perturbations associated with spiral arms are of order $|\Delta v| = 20$ km s^{-1}. Setting $\rho = 0.042$ M$_\odot$pc^{-3} (local gas density), the energy density becomes $\frac{1}{2}\rho\Delta v^2 = 6 \times 10^{-2}$ ergs cm^{-3}. Therefore, the magnetic energy density must be at least this large for magnetohydrodynamic theories to work; hence $B > 10^{-5}$ Gauss or 10^{-9} Tesla.

However, Faraday rotation and dispersion measurements of pulsars indicate a mean field of only $B \approx 2 \times 10^{-6}$ Gauss (Spitzer 1978). Since pulsars are associated with some of the strongest magnetic fields in galaxies, this is too small for the interstellar magnetic field to play a major role in spiral structure.

4.3 Driving by bars and oval distortions

In barred galaxies, the spiral arms tend to start at the end of the bar, which suggests that there may be a relation between the bar and the spiral. Consider the model investigated by Sanders and Huntley (1976). They constructed a uniform gas with no self-gravity. To start with, the gas is in circular orbits and there is a fixed central force field. An extra potential due to a rigidly rotating bar is then introduced. The gas eventually settles into a steady state exhibiting a strong trailing pattern, as shown in figure 4.2.

Note the following:
1. The gas response is spiral, but the bar forcing is not.
2. The gas response is very strong with the density enhanced by a factor of 2.5, even though the bar can be weak.

The spirality is due to the viscosity of the disk; a simulation of a disk of test particles does not yield a spiral pattern (Sanders 1977).

Figure 4.2. Steady-state simulation of the gas density distribution for the bar-driving model of the barred spiral galaxy, NGC 1300. Copyright AAS. Reproduced with permission from Roberts, Huntley and van Albada (1979).

This demonstrates that the formation of spiral arms requires only a bar and a dissipative interstellar medium. Spiral density waves are not needed. Therefore, could all unbarred galaxies possess a weak oval distortion in their centers, strong enough to drive spiral structure, but too weak to appear bar-like in observations? The answer is probably not, for two reasons:

1. Bar forcing seems unable to produce tightly wound spirals.
2. The spiral pattern is seen in old disk stars as well as young stars and gas.

Sellwood and Sparke (1988) used N-body simulations to show that bar driving is only important for the most strongly barred spiral galaxies. In addition, it is thought that bar forcing should not extend far outside the region where the bar potential is strong (Sanders and Tubbs 1980). It therefore seems that bars can induce a strong response in the gas but not in the stellar disk.

The relation between the morphology of bars and spiral arms has been discussed by Sellwood and Sparke (1988), who claimed that bars and arms may have different pattern speeds. If this were the case, then galaxies would sometimes have arms starting ahead of the bar ends, although their simulations show that this should not happened very often. They find that in 100 time-steps of their model, there are only two steps for which the ends of the bar and beginnings of the arms do not meet (see figure 4.3). Indeed, in the sample of 45 spiral galaxies studied by Seigar and James (1998), one galaxy also showed signs that the ends of the bar do not meet up with the beginnings of its spiral arms (figure 4.3).

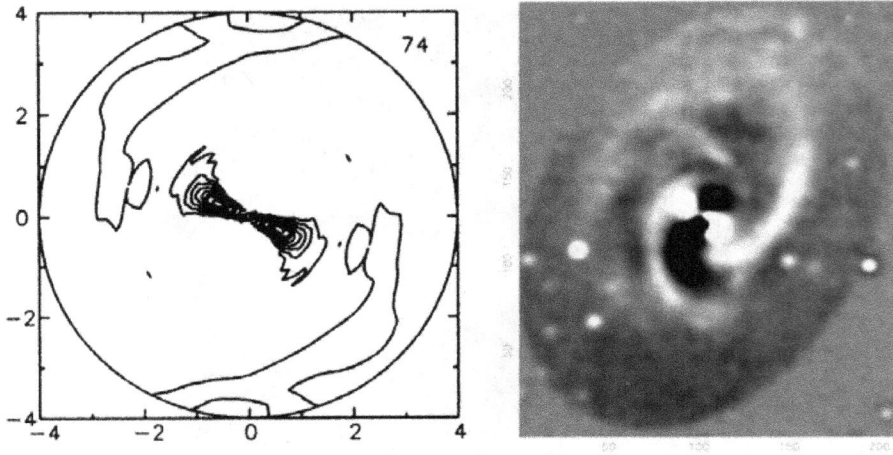

Figure 4.3. Left panel: Contours in the fractional perturbation in the density, from Sellwood and Sparke (1988), showing a time-step where the ends of the bar do not coincide with the beginnings of the spiral arms. Reproduced from Sellwood and Sparke (1988) with permission of the Royal Astronomical Society. Right panel: greyscale image of UGC 3900 after subtraction of an elliptically smoothed profile, showing that its arms start ahead of its bar, taken from Seigar and James (1998).

Suggested further reading

Athanassoula E 1980 *Astron. Astrophys.* **88** 184–93
Binney J and Tremaine S 1987 *Galactic Dynamics* (Princeton, NJ: Princeton University Press)
Contopoulos G 1985 *Nonlinear Phenomena in Physics* (New York: Springer) pp 238–54
Contopoulos G and Grosbol P 1986 *Astron. Astrophys.* **155** 11–23
Contopoulos G and Grosbol P 1988 *Astron. Astrophys.* **197** 83–90
Goldreich P and Tremaine S 1978 *Astrophys. J.* **222** 850–8
Kaufmann D E and Contopoulos G 1996 *Astron. Astrophys.* **309** 381–402
Patsis P A, Contopoulos G and Grosbol P 1991 *Astron. Astrophys.* **243** 373–80
Patsis P A, Hiotelis N, Contopoulos G and Grosbol P 1994 *Astron. Astrophy.* **286** 46–59
Patsis P A and Kaufmann D E 1999 *Astron. Astrophys.* **352** 469–78
Roberts W W, Huntley J M and van Albada G D 1979 *Astrophys. J.* **233** 67–84
Sanders R H and Huntley J M 1976 *Astrophys. J.* **209** 53–65
Sanders R H and Tubbs A D 1980 *Astrophys. J.* **235** 803–20
Schwarz M P 1984 *Mon. Not. R. Astron. Soc.* **209** 93–109
Schwarz M P 1985 *Mon. Not. R. Astron. Soc.* **212** 677–86
Seigar M S and James P A 1998 *Mon. Not. R. Astron. Soc.* **299** 685–97
Sellwood J A and Sparke L S 1988 *Mon. Not. R. Astron. Soc.* **231** 25–31
Sellwood J A 2000 *Astrophys. Space Sci.* **272** 31–43
Sparke L S and Sellwood J A 1987 *Mon. Not. R. Astron. Soc.* **255** 653–75
Spitzer L 1978 *Physical Processes in the Interstellar Medium* (New York: Wiley)
Voglis N 2003 *Mon. Not. R. Astron. Soc.* **344** 575–82

Chapter 5

Star formation in spiral galaxies

Star formation occurs in regions associated with Giant Molecular Clouds in galaxies, such as the Orion Nebula in the Milky Way Galaxy (see figure 5.1). There regions seem to align with the spiral arms in galaxies. There are two theories that try to explain how star formation may occur in spiral galaxies and they are both based on models of spiral structure. Indeed, they both make use of the concept of a density wave.

5.1 The large-scale shock scenario

The large-scale shock scenario was first proposed by Roberts (1969). In this hypothesis, the gas in a spiral galaxy settles into a quasi-stationary state, which is driven by the gravitational field of the Galaxy. The gas response can be non-linear to an imposed azimuthal sinusoidal potential, if the relative motion between the density wave and the cold interstellar medium is supersonic. This leads to the formation of a shock near the trailing edge of the spiral arms, which compresses the gas to densities at which stars can form. In observations of spiral galaxies, the shock is thought to be characterized by dust lanes seen on the trailing edges of arms, and star forming regions seen on the leading edges of arms. This is evidence that supports the large-scale shock scenario.

5.1.1 Response of gas to a density wave

The following analysis describes how shocks in density waves can lead to the formation of stars. It is based on analyses by Roberts (1969) and Binney and Tremaine (2008).

For a first approximation, we can assume that the gas moves in the gravitational field of the stars alone. The potential can therefore be written as the sum of the unperturbed axisymmetric potential, $\Phi_0(r)$, and the perturbed potential due to the stellar density wave:

$$\Phi_1(r, \phi) = \mathrm{Re}[\Phi_a(r)\exp(im\phi)] \qquad (5.1)$$

doi:10.1088/978-1-6817-4609-8ch5

Figure 5.1. Hubble Space Telescope of the Orion Nebula, a star formation region in the Orion constellation. All imaging cameras onboard the Hubble Space Telescope were used to create this image, from the ultraviolet, through the visible, to the infrared. Image courtesy of NASA/Hubble Space Telescope and the Hubble Space Telescope Orion Treasury Project.

where ϕ is the azimuthal angle in the frame rotating at Ω_p and m is the number of arms. If we assume that the wave is tightly wound, then in the solar neighborhood we have

$$\Phi_a(r) \approx F e^{ikr} \tag{5.2}$$

where F = constant. We then chose the zero point of the azimuthal angle such that F is real and we get

$$\Phi_1(r, \phi) = F\cos(kr + m\phi). \tag{5.3}$$

The interstellar clouds contain the bulk of the gas, so as a first approximation, we regard these clouds as test particles moving in the potential of the stellar disk. Their motions can then be calculated using equations of motion for a star in a weak non-axisymmetric field using the following:

$$r(t) = r_0(t) + t_1(t), \quad \phi(t) = \phi_0(t) + \phi_1(t) \tag{5.4}$$

such that

$$\Omega(r) = \sqrt{\frac{1}{r}\frac{d\Phi_0}{dr}} \tag{5.5}$$

and so the equations of motion become

$$\frac{d^2 r_1}{dt^2} + \left(\frac{d^2\Phi_0}{dr^2}\right)r_1 - 2r_0\Omega_0\frac{d\phi_1}{dt} = kF\sin(kr + m\phi) \tag{5.6}$$

$$\frac{d^2\phi_1}{dt^2} + 2\Omega_0\frac{1}{r_0}\frac{dr_1}{dt} = \frac{nF}{r_0^2}\sin(kr + m\phi). \tag{5.7}$$

Since the waves are tightly wound, the tangential force is smaller than the radial force by a factor kr_0/m. Therefore,

$$\frac{d\phi_1}{dt} + \frac{2\Omega_0 r_1}{r_0} = \text{constant} \tag{5.8}$$

and this is an expression of conservation of angular momentum. A value for r_0 can be chosen such that the right hand side is zero and ϕ_1 can therefore be eliminated from equation (5.6). In this case, ϕ is replaced by an unperturbed value $(\Omega_0 - \Omega_p)t$, thus

$$\frac{d^2 r_1}{dt^2} + \kappa_0^2 r_1 = kf\sin[k(r_0 + r_1) + m(\Omega_0 - \Omega_p)t]. \tag{5.9}$$

Imagine an endless row of gas clouds, each acting as a simple harmonic oscillator of frequency κ_0. The equilibrium position is described by the coordinate r_0 and the horizontal displacement from the equilibrium position is r. Each cloud is subject to a horizontal force per unit mass $kF\sin(kr + \omega t)$. The equation of motion is (5.9) with $\omega = m(\Omega_0 - \Omega_p)$.

Equation (5.9) must be solved numerically for the displacement $r_1(r_0, t)$. If the forcing, F, is strong enough, adjacent clouds may collide. A collision takes place if two clouds initially separated by Δr_0 occupy the same position, i.e., if $r_0 + r_1(r_0, t) = r_0 + \Delta r_0 + r_1(r_0 + \Delta r_0, t)$. Letting Δr_0 tend to zero, the condition for a collision becomes $dr_1/dr_0 < -1$.

In the linear approximation,

$$r_1 = \frac{kF}{\kappa_0^2 - \omega^2}\sin(kr_0 + \omega t) \tag{5.10}$$

and so collisions occur if

$$g = \frac{k^2 F}{|\kappa_0^2 - \omega^2|} > 1. \tag{5.11}$$

The collisions of many gas clouds correspond to the narrow regions of high-density shocked gas. Dense, narrow gaseous arms are a natural consequence of the response of gas to a density wave.

Concentration of dust also increases in these high-density areas, hence the observed dust lanes. Gas compression is accompanied by strong enhancement in magnetic field strength, since field lines are frozen into the gas, and this leads to enhanced synchrotron emission from relativistic electrons. All these phenomena are displaced from the optical arms due to the delay required for stars to form. Below is a calculation, which determines the typical thickness of a spiral arm, based upon the lifetimes of the brightest stars.

Fujimoto (1968), Roberts (1969), Shu *et al* (1972), Tosa (1973), Woodward (1975) and Nelson and Matsuda (1977) all predicted the existence of narrow spiral arms using the Lin–Shu hypothesis. The high gas density triggers rapid star formation and the lifetimes of the brightest stars are so short that they cannot drift far from their formation sites before dying. Binney and Tremaine (1987) derived the angular width of such an arm using the lifetime, t_*, of stars created in the arm. The width $\Delta\theta = |\Omega - \Omega_p| t_*$. Typically, $\Omega = 2\Omega_p$, $\Omega_p = 10$ km s^{-1} kpc^{-1} and $t_* = 2 \times 10^7$ yr for a 10M$_\odot$ star. Therefore, $\Delta\theta = 12°$, which agrees well with observations (e.g., Nakai *et al* 1994).

Johns and Nelson (1986) report a similar result to the above authors, but they also found enhanced star formation in the vicinity of corotation and a small amplitude modulation along the spiral arms with a wavelength of the order of a few kiloparsecs. This is consistent with modal theory (Bertin *et al* 1989a, b; Bertin and Lin 1996).

5.1.2 Observational evidence in favor of the large-scale shock scenario

Advances in the 1990s in this area arose from studies of the atomic and molecular gas components in spiral galaxies, through studies of HI and CO line emission. This allowed detailed mapping of both the distribution and velocity field of the gas in spiral arms of nearby galaxies, e.g., M51 (Tilanus and Allen 1991; Rand 1993; Vogel *et al* 1993; Nakai *et al* 1994; Louie *et al* 2013), M83 (Tilanus and Allen 1993; Kenney and Lord 1991; Lord and Kenney 1991), and M100 (Knapen 1993; Rand 1995). More recent studies with the Atacama Large Millimeter Array (ALMA) have been able to look at Giant Molecular Clouds (GMCs) across spiral arms and in inter-arm regions (e.g., Pan and Kuno 2017). These studies tend to find streaming velocities of gas through the spiral arms of order a few 10s of km s^{-1}, and also find offsets between the peaks of the gas density and the old stellar population in the arms and the star formation as revealed by Hα emission. Indeed, a study by Seigar and James (2002) found that Hα emission is enhanced in the vicinity of spiral arms by comparing Hα imaging and K band imaging of 20 spiral galaxies.

Some groups have tried to compare multi-waveband imaging data of several spiral galaxies in order to determine the offset between wavebands that trace star formation (e.g., the B band which traces bright, short-lived stars) and wavebands

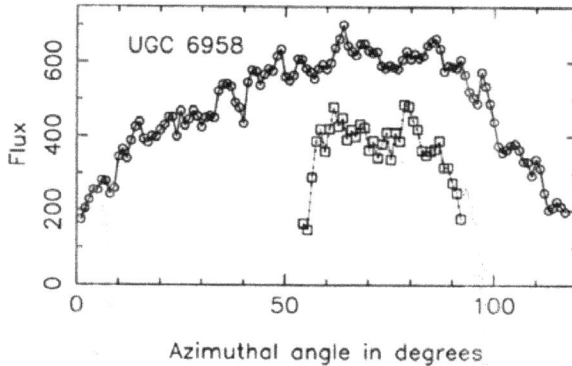

Figure 5.2. Comparison of K band cross-sectional arm profile with a B band cross-sectional arm profile for the spiral galaxy UGC 6958 from Seigar and James (1998). The B band arm profile is offset to the right (i.e., the leading) edge of the K band spiral arm. This is the expectation from density wave theory.

that trace the underlying density wave (by looking at the long-lived stars in the near-infrared in either the K band from the ground or at 3.6 μm with the Spitzer Space Telescope). Seigar and James (1998) compared the K band and B band spiral arms for three spiral galaxies and found that the blue light was consistently offset toward the leading edge of the near-infrared arms. Furthermore, the arms appeared more asymmetric in blue light than in red light (for an example see figure 5.2). Both of these results are consistent with the expectations of density wave theory. Grosbol and Patsis (1998) also found consistent offsets using visible and infrared wavebands for a total of five spiral galaxies. A more recent study by Pour Imani *et al* (2016) measured spiral arm pitch angles in visible and infrared wavebands. They found a tighter pitch angle in the B band than for the near-infrared. This result is consistent with the findings of Seigar and James (1998) and Grosbol and Patsis (1998).

All of these findings are in general agreement with the predictions of density wave theories and the large-scale shock scenario with the gas being compressed and shocked as it flows into the spiral arms, and subsequently forming stars downstream from the density peak of the spiral arms.

5.2 Stochastic self-propagating star formation

Öpik (1953) first hypothesized that a supernova explosion could trigger star formation. This was developed further as a 'spiral detonation wave' model by Mueller and Arnett (1976). This was then developed further into what is now known as self-propagating star formation (SPSF). This section describes how this idea has developed over the years into a model capable of explaining star formation in Grand Design spiral galaxies.

Mueller and Arnett (1976) used two two-dimensional disks divided up into N rings in their simulations based on SPSF. Each of the rings was further divided azimuthally into cells, such that each cell in the array had an equal area. About 1% of the cells were populated with young bright stars. One 'time unit' later, these stars would induce star formation in adjacent cells in a completely deterministic way (i.e., there was a 100% probability that star formation would occur in the adjacent cells). The possible mechanisms for inducing star formation included the following:

1. Shocks from a supernova explosion compress the gas to a density high enough for star formation to occur.
2. Ionization fronts from bright massive stars increase the gas density and induce star formation.

Mueller and Arnett (1976) showed that differential rotation of the rings (within the disk) then formed similar spiral structure to those seen in galaxies, particularly flocculent spiral galaxies with less regular spiral structure when compared to Grand Design spirals.

5.2.1 The introduction of stochasticity

Gerola and Seiden (1978) realized that a fundamental problem with the model proposed by Mueller and Arnett (1976) was that, although a bright massive star could possibly induce star formation, this would not be completely deterministic (i.e., there would be a finite probability that star formation would be induced in an adjacent cell). This is known as stochasticity. The principle behind this idea is that there is no assurance that a star will induce the creation of another massive star, because star formation is dependent on the local matter distribution and the dynamical properties of the gas. Also, massive stars only form a small percentage of the total number of stars (Salpeter 1965). The other addition included by Gerola and Seiden (1978) was to increase the size of the array so that resolution effects would not dominate the results.

Gerola and Seiden (1978) also populated 1% of their cells with 'stars' but they made the distinction that they actually meant groups of young stars, open clusters, and associations with HII regions. Again, one 'time unit' later, these 'stars' could possibly induce star formation in an adjacent cell. The probability of induced star formation was P_{st}.

Once a star is created in a specific cell it is harder to produce another star in that cell for a time period τ_r. This is due to the fact that a lot of the gas in that cell has already been used to create a 'star'. In this model, stars are also spontaneously created at random with probability P_{sp} but $P_{sp} \ll P_{st}$.

The probability of creating a star in a cell in which a star has recently been created changes smoothly over τ_r. The function describing this is as follows:

$$P = P_{st}\frac{\tau_a}{\tau_r} \tag{5.12}$$

where P is the probability of creating a star and τ_a is the age of the star last created in that cell. Equation (5.12) holds when $\tau_a < \tau_r$, otherwise $P = P_{st}$.

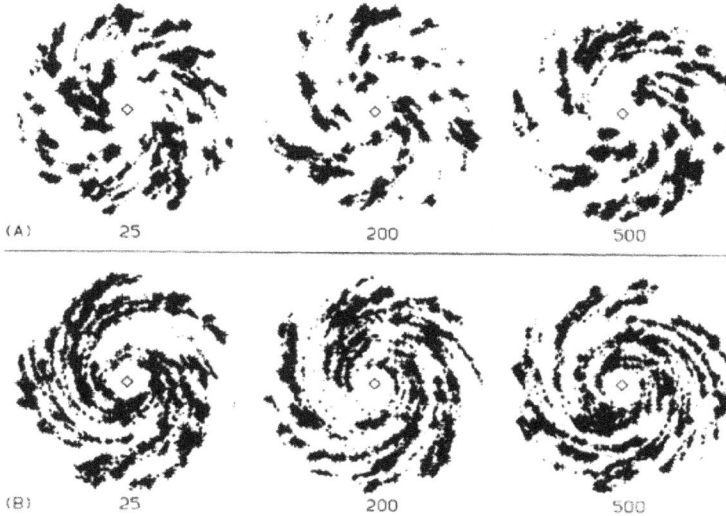

Figure 5.3. Evolution of two model galaxies. Copyright AAS. Reproduced with permission from Gerola and Seiden (1978). The top three examples (A) are for the rotation curve of M101, and the bottom three (B) are for M81. The number below each figure is its age in time steps. Each time step equals 15 million years. The outside rim completes one revolution in about 32 time steps.

Given a galactic radius G_R and N rings, then the size of a cell is G_R/N. A time step is defined as the time it takes a supernova remnant or an ionization front from a massive star to reach an adjacent cell, i.e.,

$$\delta t = \frac{G_R}{N V_1} \tag{5.13}$$

where V_1 is the velocity of propagation of the supernova or ionization front.

The parameter τ_r is defined as the length of time it takes gas to return to a cell, which has had a supernova event. Gas can return to a cell by one of two mechanisms:

1. A molecular cloud moves into a cell.
2. Gas diffuses into the cell from other high-density regions. Gerola and Seiden (1978) assumed that diffusion was the dominant factor in returning gas to a cell. Therefore, gas returns to a cell by diffusion at the velocity of sound, V_s, of the interstellar gas so that $\tau_r = V_1/V_s$. It can be shown that the results are not very sensitive to τ_r.

In figure 5.3, stars up to a lifetime of 10 time steps are plotted. This is equivalent to 150 million years. The central eight rings of the array are kept devoid of stars as Gerola and Seiden (1978) did not take the galactic bulge into consideration.

This model of stochastic self-propagating star formation (SSPSF) is able to produce large-scale spiral structure over a period of many galactic rotations. The morphological type depends only the rotation curve. As the model evolves, the Hubble type remains constant.

Gerola and Seiden (1978) also found that there is a critical range of values for P_{st} for which SSPSF occurs. If P_{st} is too high, the Galaxy undergoes an explosion of star formation, whereas if P_{st} is too low, the Galaxy completely ceases all activity.

5.2.2 Further developments of SSPSF

Seiden and Gerola (1982) used an enhanced model for SSPSF. They filled their grid with 'active' and 'inactive' gas. If star formation occurs within a cell, the gas becomes 'inactive', and over some time period the gas eventually becomes 'active' again. The probability of star formation occurring in any given cell is proportional to the 'active' gas density within that cell. Seiden (1983) identified 'active' with neutral hydrogen and the 'inactive' gas with molecular hydrogen. A study by Jungwiert and Palous (1994) incorporated an anisotropic probability distribution to represent differential shearing of the material swept up by supernova shocks. However, their simulation made no attempt to model the structure of the interstellar medium. It still proceeded on a grid with cells of either 'active' or 'inactive' gas.

A study by Sleath and Alexander (1995, 1996) attempted to model the interstellar medium in a propagating star formation model, which also incorporated spiral density wave theory. They made use of observations by Elmegreen and Elmegreen (1986), who had shown that the enhancement of spiral arms above the disk was not strong enough for density waves to be responsible for star formation, but density waves are important for organizing of the interstellar medium and stars, and for concentrating new star formation regions along the spiral arms (Elmegreen 1993).

The model by Sleath and Alexander (1995) used a particulate representation for gas clouds and stars (massless test particles). This allows for easy modeling of the dynamics of the interstellar medium and easy modeling of the effect of the spiral density wave through a realistic potential. They allow for cloud–cloud collisions, but the particles are not self-gravitating. Their model reproduced the entire morphological range of Hubble classifications realistically.

Their model consisted of three main components: (i) a diffuse gaseous component, (ii) gas clouds, and (iii) star clusters and associations. The clouds and stars were test particles moving in a galactic potential with axisymmetric and spiral parts. The total number of cloud particles was fixed at 32 000, and the number of star particles was allowed to vary. Gas clouds accrete from diffuse gas. Therefore, any structure is believed to be that of a molecular hydrogen core surrounded by an HI halo (Wannier *et al* 1983; Elmegreen 1985).

Sleath and Alexander (1995) express the probability of star formation to be induced in a cell adjacent to a cell populated by a star capable of inducing star formation as a function of the cloud mass, i.e.,

$$P_{st} = \frac{M_i}{M_{st}} \tag{5.14}$$

where M_i is the cloud mass and M_{st} is a scaling mass controlling the simulated star formation.

In a similar way, the probability for spontaneous star formation is

$$P_{sp} = \frac{M_i}{M_{sp}} \tag{5.15}$$

where M_{sp} is determines the rate of spontaneous star formation. However, $M_{st} \ll M_{sp}$ by about six orders of magnitude and so SSPSF is the dominant mechanism for star formation.

When star formation occurs, a gas cloud is disrupted and its mass is reduced from M_i to εM_i (i.e., a mass of $(1 - \varepsilon)M_i$ becomes locked in stars). Typically, $\varepsilon = 10^{-3}$ and ε is the factor by which the cloud was disrupted. As a consequence of these conditions, the rate at which supernovae occur is equal to the cluster formation rate, ψ.

The stars and clouds orbit as test masses in a galactic potential plus a spiral component. The galactic potential consists of three components as follows:

(i) The central bulge, given by the potential

$$\phi_1(r, z) = \frac{-M_1}{\left(r^2 + z^2 + b_1^2\right)^{0.5}}. \tag{5.16}$$

(ii) The disk component, given by the potential

$$\phi_2(r, z) = \frac{-M_2}{\left(r^2 + \left[a_2 + \left(z^2 + b_2^2\right)^{0.5}\right]^2\right)^{0.5}}. \tag{5.17}$$

(iii) The spherical halo

$$\phi_3(R) = -\frac{M_3 R^{1.02}}{a_3^{2.02}}\left[1 + \left(\frac{R}{a_3}\right)^{1.02}\right]$$

$$-\frac{M_3}{1.02 a_3}\left(\frac{-1.02}{\left[1 + \left(\frac{R}{a_3}\right)^{1.02}\right]} + \ln\left[1 + \left(\frac{R}{a_3}\right)^{1.02}\right]\right). \tag{5.18}$$

where $R = \sqrt{r^2 + z^2}$; a_2, a_3, b_1, b_2, M_1, M_2, and M_3 are determined by the rotation curve and the orbits of high velocity stars.

Superimposed on the galactic potential, the model included a logarithmic spiral potential as follows:

$$\phi_4 = \frac{-Ar^2}{\left(a_4^2 + r^2 + z^2\right)^2}\cos[n\theta - n\Omega_p t + \chi(r)] \tag{5.19}$$

where

$$\chi(r) = \frac{\ln\left[a + \left(\dfrac{r}{r_0}\right)^p\right]}{p \, \tan i_0} \tag{5.20}$$

and n is the number of arms, Ω_p is the pattern speed, $\chi(r)$ is the spiral shape function, i_0 is the pitch angle, and p and r_0 determine the barred qualities of the potential.

Consider the star formation rate (SFR) or the cluster formation rate (CFR). The velocity dispersion of the gas clouds ≈ 7 km s^{-1} and $M_{st} \approx 1.2 \times 10^6$ M$_\odot$. The cluster formation rate is $\psi = 4.0 \times 10^{-5}$ yr^{-1}. Taking the galactic radius to be 10 kpc, the CFR/unit area $= 1.3 \times 10^{-7}$ kpc^{-2} yr^{-1} (Sleath and Alexander 1995). The observed value is $(2.5 \pm 1.0) \times 10^{-7}$ kpc^{-2} yr^{-1} (Elmegreen and Clemens 1985).

There are three groups of control parameters in the Sleath and Alexander (1995) model. These are:

1. The star formation parameters, M_{st}, M_{sp}, and ε.
2. The cloud velocity dispersion, v_{disp}.
3. The form of the spiral potential.

The most important of these is M_{st}. The cluster formation rate is $\psi \propto M_{st}^{-0.309 \pm 0.060}$. As M_{st} increases, the probability of triggering star formation in a cloud decreases and therefore, the star formation rate is decreased. Also, $\psi \propto \varepsilon^{0.019 \pm 0.005}$ and $\psi \propto M_{sp}^{-0.005 \pm 0.001}$ (Sleath and Alexander 1995).

5.2.3 The Schmidt–Kennicutt law and SSPSF

The Schmidt–Kennicutt law says that the galactic star formation rate has a simple power law dependence on the gas density (Schmidt 1959, 1963). The original formulation was in terms of the HI gas volume density:

$$\frac{d\rho_*}{dt} \propto \rho_{HI}^n \tag{5.21}$$

where $d\rho_*/dt$ is the rate of change of the mass density of stars (i.e., the star formation rate), ρ_{HI} is the HI gas density, and generally $n = 1 - 2$.

Observational determinations of the Schmidt–Kennicutt law index, n, are based on surface density measurements, as volume densities cannot be measured from observations, i.e.,

$$\frac{d\sigma_*}{dt} \propto \sigma_{tot}^N. \tag{5.22}$$

In a study by Kennicutt (1998), disk averaged star formation rates and gas densities for a combined sample of 61 normal spiral galaxies and 36 starburst galaxies were analyzed to find a best fitting Schmidt–Kennicutt law index of $N = 1.40 \pm 0.15$.

The Sleath and Alexander (1995) model for SSPSF predicts a standard Schmidt–Kennicutt law. To convert the cluster formation rate into the total star formation rate, one must multiply ψ by a characteristic mass. Sleath and Alexander (1995) used the median cloud mass, M_{med}, such that

$$\frac{d\rho_*}{dt}V \propto (1 - \varepsilon)\psi M_{med} \tag{5.23}$$

where V is the volume of the Galaxy and $(1 - \varepsilon)$ is the amount of cloud material converted into stars. Also, note that:

$$\psi \propto M_{st}^{-0.309\pm0.006} \tag{5.24}$$

and that

$$M_{med} \propto M_{st}^{0.955\pm0.007} \tag{5.25}$$

and therefore

$$\frac{d\rho_*}{dt}V \propto M_{st}^{0.646\pm0.009} \propto M_{tot}^{1.65\pm0.04} \tag{5.26}$$

where M_{tot} is the total cloud mass. Using $\rho_{gas} = M_{tot}/V$, the following equation is derived:

$$\frac{d\rho_*}{dt} \propto \rho_{gas}^{1.65\pm0.04} V^{1.65\pm0.04} \propto \rho_{gas}^{1.65\pm0.04} D^{1.30\pm0.08} \tag{5.27}$$

where $V \propto D^2$

Sleath and Alexander (1995) tested their formulation of the Schmidt–Kennicutt law with the observational data of Young *et al* (1989). This showed that the far-infrared luminosity was a good measure of star formation rate and hence from equation (5.26), we can derive

$$L_{IR} \propto \frac{d\rho_*}{dt}V \propto M_{tot}^{1.65}. \tag{5.28}$$

Sleath and Alexander (1995) found that the data were well fitted by a power law index of 1.4 ± 0.4, which is very similar to the observational value found by Kennicutt (1998).

5.3 Summary

Density wave theories are best at describing Grand Design spiral galaxies in detail. However, a different mechanism is necessary to describe the short spokes and spirals seen in flocculent spiral galaxies. The differences between Grand Design and flocculent spiral galaxies can be seen in figure 5.4.

The predictions of spiral density wave theories (particularly modal theory) and the large-scale shock scenario do a great job at describing the structure and the observed star formation in Grand Design spirals such as M81 as shown in figure 5.4. However, the flocculent structure observed in galaxies like NGC 4414 in figure 5.4 turn out to be better described by SSPSF, both in terms of their structure and where

Figure 5.4. Left panel: a Spitzer Space Telescope image of M81, a Grand Design spiral galaxy, showing a regular two-armed spiral pattern. Image courtesy of NASA/JPL. Right panel: NGC 4414 is a flocculent spiral galaxy, without a well-defined two-armed spiral pattern. Instead it has several short spokes that make up its spiral pattern. Image courtesy of NASA/STScI/AURA.

their star formation occurs. The earlier formulations of SSPSF are great at producing the short spokes of star formation as seen in NGC 4414. Quite often, near-infrared observations of flocculent spiral galaxies reveal that an underlying density wave still exists in galaxies that appear flocculent in visible light (Thornley 1996; Seigar *et al* 2003). This gives some credence to the model of SSPSF described by Sleath and Alexander (1995, 1996).

Suggested further reading

Bertin G, Lin C C, Lowe S A and Thurstans R P 1989a *Astrophys. J.* **338** 78–103

Bertin G, Lin C C, Lowe S A and Thurstans R P 1989b *Astrophys. J.* **338** 104–20

Bertin G and Lin C C 1996 *Spiral Structure in Galaxies: A Density Wave Theory* (Cambridge, MA: MIT Press)

Binney J and Tremaine S 1987 *Galactic Dynamics* (Princeton, NJ: Princeton University Press)

Binney J and Tremaine S 2008 *Galactic Dynamics* 2nd edn (Princeton, NJ: Princeton University Press)

Elmgreen B G 1985 *Phys. Scripta* **T11** 48–52

Elmgreen B G 1985 *Birth and Infancy of Stars* ed R Lucas, A Omont and R Stora (Amsterdam: Elsevier) pp 215–54

Elmegreen B G and Clemens C 1985 *Astrophys. J.* **294** 523–32

Elmegreen B G and Elmegreen D M 1986 *Astrophys. J.* **311** 554–62

Elmegreen D M 1993 *Star Formation, Galaxies and the Interstellar Medium* (Cambridge: Cambridge University Press) pp 108–16

Fujimoto M 1968 *Astrophys. J.* **152** 391–404

Gerola H and Seiden P E 1978 *Astrophys. J.* **223** 129–35

Grosbol P J and Patsis P A 1998 *Astron. Astrophys.* **336** 840–54

Johns T C and Nelson A H 1986 *Mon. Not. R. Astron. Soc.* **220** 165–83

Jungwiert B and Palous J 1994 *Astron. Astrophys.* **287** 55–67

Kenney J D P and Lord S D 1991 *Astrophys. J.* **381** 118–29

Kennicutt R C 1998 *Astrophys. J.* **498** 541–52

Knapen J H 1993 *Publ. Astron. Soc. Pac.* **105** 323

Lord S D and Kenney J D P 1991 *Astrophys. J.* **381** 130–6

Louie M, Koda J and Egusa F 2013 *Astrophys. J.* **763** 94

Mueller M W and Arnett W D 1976 *Astrophys. J.* **210** 670–8

Nakai N, Kuno N, Handa T and Sofue Y 1994 *Publ. Astron. Soc. Japan* **46** 527–38

Nelson A H and Matsuda T 1977 *Mon. Not. R. Astron. Soc.* **179** 663–70

Opik E J 1953 *Irish Astron. J.* **2** 219–33

Pan H-A and Kuno N 2017 *Astrophys. J.* **839** 133

Pour-Imani H, Kennefick D, Kennefick J, Davis B L, Shields D W and Shameer Abdeen M 2016
   Astrophys. J. Lett. **827** L2

Rand R J 1993 *Astrophys. J.* **410** 68–80

Rand R J 1995 *Astrophys. J.* **109** 2444–59

Roberts W W 1969 *Astrophys. J.* **158** 123–43

Salpeter E E 1965 *The Structure and Evolution of Galaxies* (London: Wiley) pp 71–82

Schmidt M 1959 *Astrophys. J.* **129** 243—58

Schmidt M 1963 *Astrophys. J.* **137** 758—69

Seiden P E and Gerola H 1982 *Fundam. Cosm. Phys.* **7** 241–311

Seiden P E 1983 *Astrophys. J.* **266** 555–61

Seigar M S and James P A 1998 *Mon. Not. R. Astron. Soc.* **299** 685–98

Seigar M S and James P A 2002 *Mon. Not. R. Astron. Soc.* **337** 1113–7

Seigar M S, Chorney N E and James P A 2003 *Mon. Not. R. Astron. Soc.* **342** 1–7

Shu F H, Milione V, Gebel W, Yuan C, Goldsmith D W and Roberts W W 1972 *Astrophys. J.*
   173 557–92

Sleath J P and Alexander P 1995 *Mon. Not. R. Astron. Soc.* **275** 507–14

Sleath J P and Alexander P 1996 *Mon. Not. R. Astron. Soc.* **283** 358–66

Thornley M D 1996 *Astrophys. J. Lett.* **469** L45–8

Tilanus R P J and Allen R J 1991 *Astron. Astrophys.* **244** 8–26

Tilanus R P J and Allen R J 1993 *Astron. Astrophys.* **274** 707–29

Tosa M 1973 *Publ. Astron. Soc. Japan* **25** 191–205

Vogel S N, Rand R J, Gruendl R A and Tueben P J 1993 *Publ. Astron. Soc. Pac.* **105** 666–9

Wannier P G, Lichten S M and Morris M 1983 *Astrophys. J.* **268** 727–38

Woodward P R 1975 *Astrophys. J.* **195** 61–73

Young J S, Xie S, Kenney J D P and Rice W L 1989 *Astrophys. J. Suppl.* **70** 699–722

Chapter 6

Spiral structure and its connection with black holes and dark matter

The idea that the morphology of spiral arms is related to the central mass concentration in spiral galaxies was first addressed by the original density wave paper (Lin and Shu 1964), although they related the spiral pitch angle to the fraction of mass in the disk (i.e., the reciprocal of central mass concentration). This predicts that galaxies with a greater mass concentration (i.e., a larger bulge, which in turn would suggest a larger central black hole—see figure 6.1 for an artist's impression)

Figure 6.1. An artist's impression of a supermassive black hole in the center of the Milky Way Galaxy. Image courtesy of NASA/STScI/AURA.

doi:10.1088/978-1-6817-4609-8ch6

should have more tightly wound spiral arms, a central feature of the Hubble sequence. Lin and Shu (1964) derived the following expression for the locus of the spiral pattern:

$$n(\theta - \theta_0) = -\int_{r_0}^{r} \frac{\left[\kappa^2 + \omega_i^2 + (\omega_r - n\Omega)^2\right]}{(2\pi G\mu_0)} dr \qquad (6.1)$$

where θ is the azimuthal angle, κ is the epicyclic frequency, μ_0 is the fraction of mass in the disk, Ω is the pattern speed, and ω_r and ω_i are the real and imaginary parts, respectively, of the material speed.

The pitch angle, i, is given by

$$\cot(i) = r\frac{d\theta}{dr}. \qquad (6.2)$$

Differentiating equation (6.1) and substituting from equation (6.2) gives,

$$\cot(i) = -\frac{r}{n}\frac{\left[\kappa^2 + \omega_i^2 + (\omega_r - n\Omega)^2\right]}{2\pi G\mu_0}. \qquad (6.3)$$

Equation (6.3) contains the desired result, that galaxies with more dominant disks should have spiral structure that is more loosely wound. Roberts *et al* (1975) converted this into a correlation between pitch angle and Hubble type, which they assumed to be essentially determined by the bulge-to-disk ratio. They calculated the theoretical pitch angle from the bulge-to-disk ratios of a sample of galaxies and plotted this versus Hubble type. Roberts *et al* (1975) claimed that although the pitch angle has some dependence on the dynamical properties of the disk, i.e., κ, ω_r and ω_i in equation (6.3), its dependence upon the fraction of mass in the disk is stronger. The dynamical properties, especially the material speed, seem to correlate more strongly with the luminosity class (van den Bergh 1960) than with Hubble type. From and observational standpoint, investigations of whether pitch angle correlates well with Hubble type produce weak correlations at best, with a lot of scatter (Kennicutt 1981; Seigar and James 1998). However, when Lin and Shu (1964) first formulated their density wave theory, little or nothing was known about dark matter in galaxies. It took the observations of rotation curves by Vera Rubin and her collaborators to highlight that dark matter was necessary to stabilize the dynamics of disk galaxies. It could be, that just looking at the bulge-to-disk ratio (or its reciprocal) is missing part of the picture, and that the dynamical mass of the bulge is a more fundamental parameter in density wave theory.

6.1 The connection with dark matter

In a series of publications by Seigar *et al* (2004, 2005, 2006, 2014) it was shown that there is a good correlation between spiral arm pitch angle and the rate of shear in disk galaxies. The most up-to-date version of this correlation can be seen in figure 6.2. To understand this relation, the rate of shear first has to be defined, and it is defined as

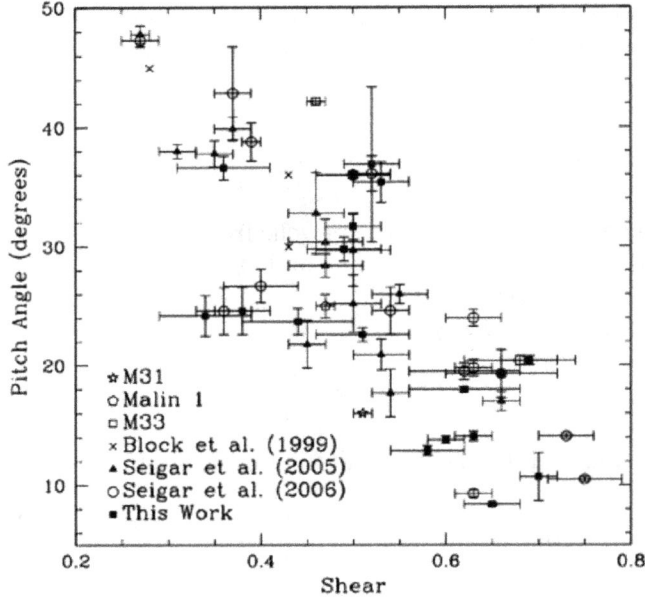

Figure 6.2. Spiral arm pitch angle versus rotation curve shear rate from Seigar *et al* (2014), showing a correlation. The filled triangles represent galaxies with data measured by Seigar *et al* (2005), the open circles are galaxies from Seigar *et al* (2006), the crosses are galaxies from Block *et al* (1999), the open square represents data for M33 (Seigar 2011), the open pentagon represents data for Malin 1 (Seigar 2008), the star represents data for M31 (Seigar *et al* 2008a), and the filled squares represent the data from Seigar *et al* (2014).

$$S = \frac{A}{\omega} = \frac{1}{2}\left(1 - \frac{R\,dV}{V\,dR}\right) \qquad (6.4)$$

where A is the first Oort constant, ω is the angular velocity, and V is the velocity at radius R.

The rate of shear depends upon the shape of the rotation curve. For a rotation curve that remains flat, $S = 0.5$; for a falling rotation curve, $S > 0.5$; and for a continually rising rotation curve, $S < 0.5$. As the shape of the rotation curve depends on the mass distribution, the shear rate at any given position depends on the mass within that radius or the central mass concentration. As a result, the spiral arm pitch angle is dependent upon the central mass concentration, and this is consistent with the expectations of most spiral density wave models (e.g., Bertin *et al* 1989a, 1989b; Bertin 1991, 1993, 1996; Bertin and Lin 1996; Fuchs 1991, 2000). Indeed, because the pitch angle depends on the enclosed mass (as a function of radius), it is actually expected to vary slightly as one moves outwards in the disk. Davis *et al* (2015), showed that pitch angle follows the following dependency,

$$\tan|i| \propto \frac{\sigma_0}{M_{\text{bulge}}} \qquad (6.5)$$

where i is the pitch angle, M_0 is the bulge mass, and σ_0 is the gas density in the disk.

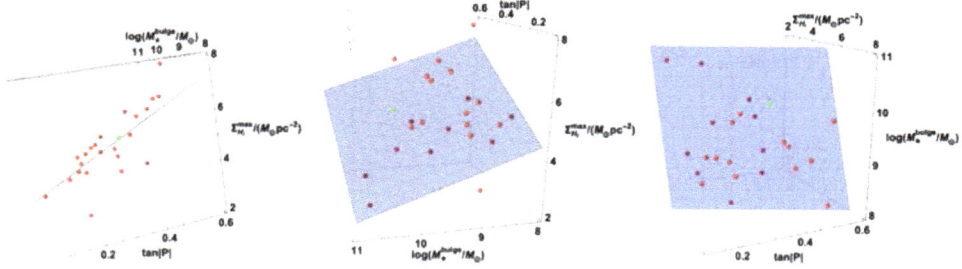

Figure 6.3. Three-dimensional plot of the plan defined by fitting to equation (6.5). The translucent blue mesh shows the best fitting plane along with plotted point of 24 galaxies studied by Davis *et al* (2015). The 24 galaxies are depicted by red spheres with the Milky Way in green. The points appear slightly darker when they are projected behind the blue plane. Left panel: the view has been oriented parallel to the plane. Middle panel: the view has been oriented at an angle sufficient to view the face of the plane. Right panel: the view has been projected along an orthogonal vector above the plane.

The affect is that spiral arm pitch angle is often seen to tighten as a function of radius in spiral galaxies, and this is often seen (Davis *et al* 2012). Davis *et al* (2015) go on to show that a strong relationship exists between the mean pitch angle in spiral galaxies, the HI gas density in the disks of galaxies, and the stellar mass of their bulges, thus defining a fundamental plane for spiral structure, and strongly supporting the modal theory of spiral structure in disk galaxies (see figure 6.3).

The relationship between spiral arm pitch angle and shear rate (which can be used as a proxy for central mass concentration) as shown in figure 6.2, shows a much better correlation than the relationship between pitch angle and bulge-to-disk ratio. This may be because, the central mass concentration, as determined by shear, is actually a dynamical mass (and therefore includes dark matter, if only a small amount), whereas the bulge mass (or light, assuming light traces stellar mass) in the bulge-to-disk ratio, is a measure of the bulge stellar mass only. This could be the reason for the tightness of the relation seen in figure 6.2.

With this in mind, Seigar *et al* (2006, 2014) developed and applied a method for using the relationship between pitch angle and shear to put constraints on the distribution of mass in dark matter halos in spiral galaxies. The main conclusion from Seigar *et al* (2014; shown in figure 6.4) presents initial evidence of a relationship between spiral arm pitch angle and dark matter halo concentration (as defined by Navarro *et al* 1996, 1997). While, this correlation is weak, more data is needed to study this relationship in more detail. In this relationship, the dark matter halo concentration is defined using the Navarro *et al* (1997) density profile for dark matter halos, which is given by

$$\rho(r) = \frac{\rho_s}{(r/r_s)(1 + r/r_s)^2} \tag{6.6}$$

where r_s is a characteristic inner radius and ρ_s is the density at that radius. This is a two-parameter function and can be completely specified by chosing two independent

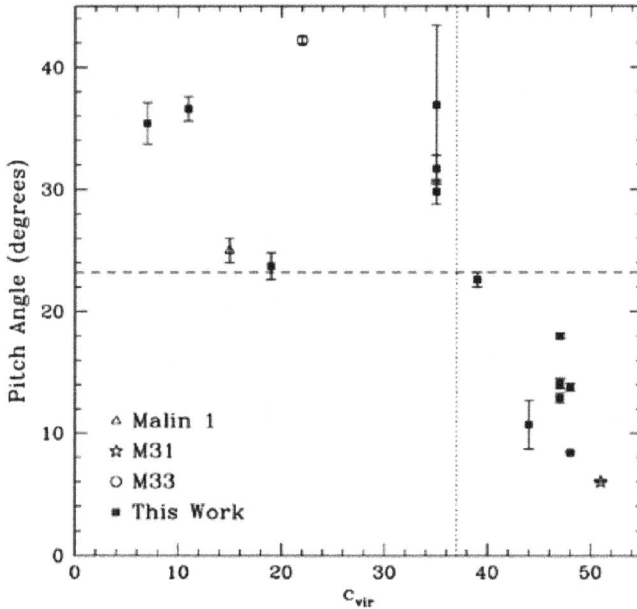

Figure 6.4. Spiral arm pitch angle versus dark matter halo concentration from Seigar *et al* (2014). The open triangle represents Malin 1 (Seigar 2008), the star represents M33 (Seigar 2011), and the filled squares represent the sample studied by Seigar *et al* (2014).

parameters, such as the virial mass, M_{vir}, and halo concentration, c_{vir}, which is defined as,

$$c_{vir} = \frac{R_{vir}}{r_s} \tag{6.7}$$

where R_{vir} is the virial radius (Bullock *et al* 2001).

The advantage of being able to use spiral structure as an indicator of the concentration (or distribution) of dark matter in galaxies, is that spiral structure can be seen in galaxies at redshifts above $z = 1$ (Elmegreen *et al* 2004), which is equivalent to a look-back time of 8 billion years (i.e., a distance of 8 billion light-years), which is less than half the current age of the Universe (estimated to be ~13.8 billion years). As a result, spiral structure (or more precisely, spiral arm pitch angle) can be used as a proxy for mass buildup in galaxies over time. Indeed, figure 6.5 shows application of this method to a spiral galaxy at a redshift of $z = 1.0$, TKRS 2012 (Weiner *et al* 2006). This galaxy is in the GOODS North field, and this analysis was performed by the author of this book, and it is being published for the first time here.

It is important to note that cosmologists expect galaxies to become more centrally concentrated over time as mass falls inwards. However, measuring this in observed galaxies has been problematic until now. The method of using spiral arm pitch angle as a proxy for mass concentration, has, for the first time, given astronomers the tools to measure mass concentrations in galaxies at intermediate redshifts.

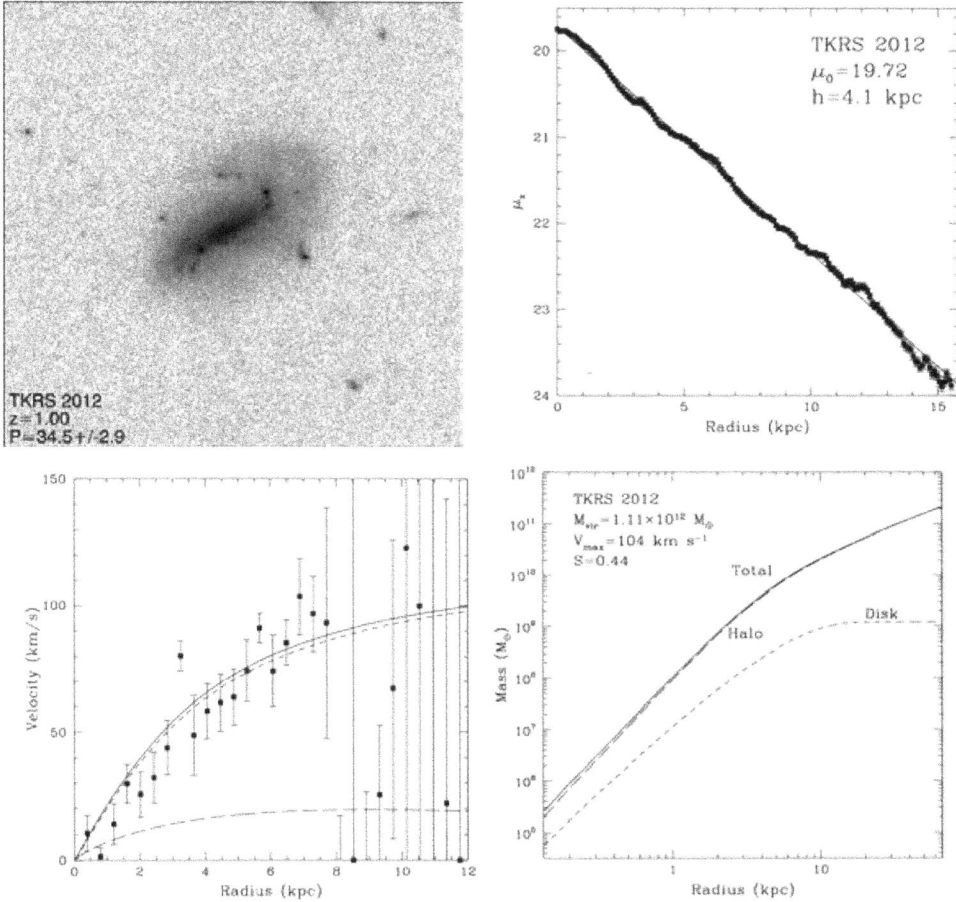

Figure 6.5. The results of modeling a redshift $z = 1.0$ galaxy, TKRS 2012, are shown. Top left: HST z-band image of TKRS 2012, a galaxy in the GOODS-N field and a redshift, $z = 1.0$. Spiral structure is clearly shown. Top right: z-band surface brightness profile of TKRS 2012, showing that it is a pure disk galaxy. Bottom left: rotation curve data from Weiner *et al* (2006) with the best-fit pseudo-isothermal model overlaid (short-dashed line) along with the disk contribution (long-dashed line) and the total of the halo+disk (solid line). Bottom right: enclosed mass as a function of radius, decomposed into disk (short-dashed line) and halo (long-dashed) components (the solid line represents the total mass of the disk+halo).

6.2 The connection with supermassive black holes

The relationship between rotation curve shear and spiral arm pitch angle, shown in figure 6.2, suggests that the tightness of spiral structure is most closely related to the central mass concentration. In most spiral galaxies, the central mass will be dominated by the bulge mass. In essence, this relationship is indicative of the Hubble sequence, where galaxies with large bulges (i.e., Sa-type galaxies) have tightly wound arms, and galaxies with small bulges (i.e., Sc-type galaxies) have loosely wound arms. Since the launch of the Hubble Space Telescope, it has also become evident that almost every galaxy in the Universe harbors a supermassive black hole (SMBH) at its nucleus, with a mass in the range of millions to billions of solar masses (e.g., Magorrian *et al* 1998). The masses of these SMBHs are related to the luminosity (Magorrian *et al* 1998) and stellar mass of the central bulge (Haring and Rix 2004) in disk galaxies. Given, this and the relationship between shear and spiral pitch angle shown in figure 6.2, we would also expect to find a relationship between SMBH mass and spiral arm pitch angle. Indeed, this relationship between SMBH mass and pitch angle was discovered by Seigar *et al* (2008b) and later confirmed by Berrier *et al* (2013) and the relationship is shown in figure 6.6.

This relationship is important for several reasons. Firstly, it is relatively cheap in terms of telescope time. Measuring SMBH masses has usually relied on a determination of central velocities of stars or gas, a measurement that is made using the Doppler shifts of absorption or emission lines observed in galaxy spectra. Spectra are expensive to obtain in terms of telescope time. The exposures times are long. However, measurement of spiral arm pitch angle requires just an image of a galaxy, which is a lot cheaper in terms of telescope time because the exposure times are relatively short. Also, spiral arm pitch angle can be measured for higher redshift galaxies, at least up to redshifts of $z = 1.0$ and maybe even further (Elmegreen *et al* 2004). This allows for an easy determination of SMBH masses in galaxies at these redshifts.

The relationship between spiral arm pitch angle has allowed scientists to determine a local black hole mass function (BHMF) for spiral galaxies (see Davis *et al* 2014). Following this work, a BHMF was then determined for all galaxy types in the same local volume, using just galaxy images (Mutlu-Pakdil *et al* 2016). In this case, the black hole masses for spiral galaxies were estimated from their spiral pitch

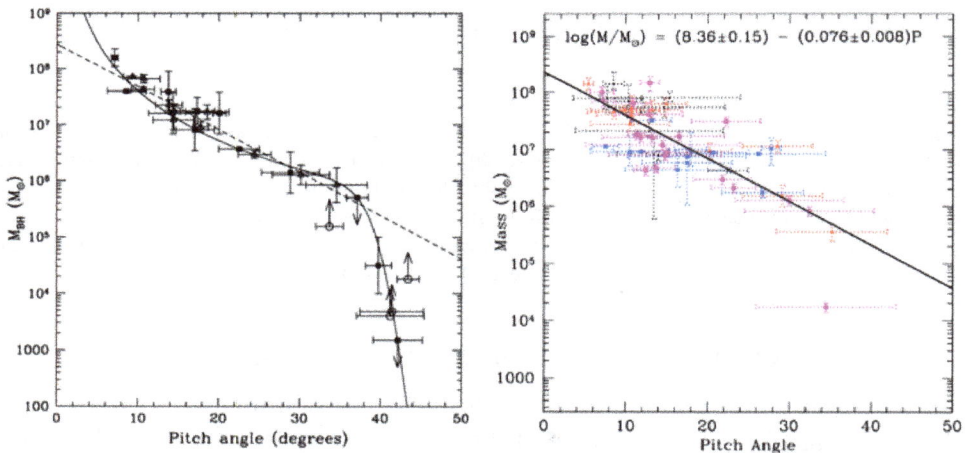

Figure 6.6. Supermassive black hole mass as a function of spiral arm pitch angle. Left panel: the initial discovery of this relationship by Seigar *et al* (2008b). Right panel: the follow-up result by Berrier *et al* (2013), including all directly measured black hole masses and black hole masses determined via central stellar or gaseous velocity dispersion. The linear versions of the relationship in both studies are consistent with each other.

angles. For determination of black hole masses in elliptical and lenticular galaxies, Mutlu-Pakdil *et al* (2016) used the relationship between galaxy light concentration and black hole mass (which is easily determined for elliptical galaxies; Graham and Driver 2007), which can also be determined just from images of galaxies. The resulting BHMF for these studies can be seen in figure 6.7.

It turns out that the best indicator of SMBH mass in spiral galaxies is spiral arm pitch angle. The scatter in this relation is smaller than for any other relation (Berrier *et al* 2013), including that of stellar velocity dispersion (Gultekin *et al* 2009). As a result, the BHMFs shown in figure 6.7 are the most reliable versions, particularly for late-type (i.e., spiral) galaxies.

The fact that SMBH masses correlate so well with spiral arm pitch angles is an example of an SMBH scaling relation beyond the bulge. Until recently, it was clear that SMBH mass scaled well with several properties of their host bulges (the bulge mass as shown by Haring and Rix 2004, or the central velocity dispersion as shown by Gebardt *et al* 2000 and Ferrarese and Merritt 2000). More recent discoveries, such as the SMBH mass versus pitch angle relation, suggest that SMBHs and spiral arms are both regulated by a more global phenomenon, which probably also regulates the properties of the bulge. It has been suggested that the regulating factor is the dark matter halo, and this makes sense in light of the fact that pitch angle correlates well with both mass concentration and SMBH mass.

To investigate this idea, Mutlu-Pakdil *et al* (2017) looked at the results of a theoretical cosmological simulation known as Illustris (Genel *et al* 2014;

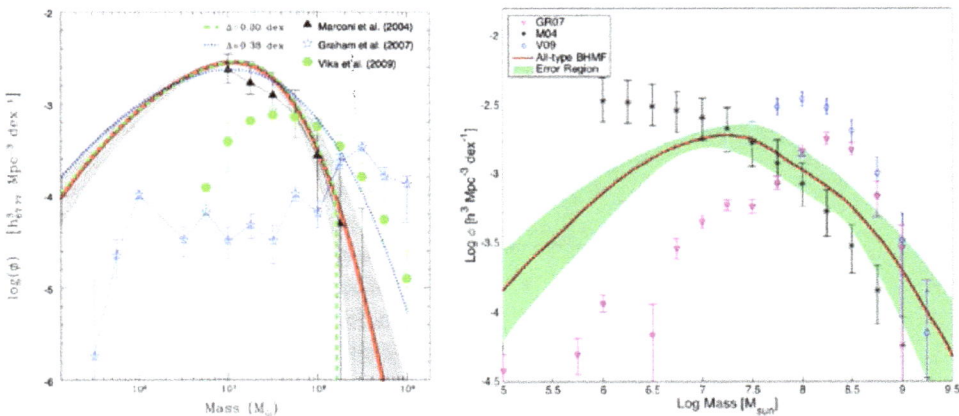

Figure 6.7. Left panel: the BHMF determined by Davis *et al* (2014) for late-type (spiral) galaxies is shown in red line with the 1σ uncertainties indicated by the grey shaded area. The BHMF shown here is derived using spiral arm pitch angles. This BHMF is compared with those of Marconi *et al* (2004), depicted by black triangles, Graham *et al* (2007) depicted by blue stars, and Vika *et al* (2009) depicted by green hexagons. Right panel: the BHMF for all type galaxies determined by Mutlu-Pakdil *et al* (2016) is shown in red with the 1σ uncertainties indicated by the green shaded area. The BHMF shown here is derived using spiral arm pitch angles for late-type galaxies and light concentration (or Sersic index) for early-type galaxies. This BHMF is compared with those of Graham *et al* (2007) depicted by pink triangles, Marconi *et al* (2004) depicted by black stars, and Vika *et al* (2009) depicted by blue circles.

Vogelsberger *et al* 2014). Simulated images of spiral galaxies were taken from the Illustris simulation and their pitch angles were measured directly from the simulated images. The SMBH masses were reported by the Illustris team, so Mutlu-Pakdil *et al* (2017) used these to make sure that the Illustris simulation reproduced the same relation between SMBH mass and pitch angle as that observed in real galaxies. The result of this comparison is shown in figure 6.8, which shows the measured pitch angles and SMBH masses from the Illustris simulations and compares the resulting correlation with that from observed galaxies in the local universe (Berrier *et al* 2013). The comparison shows that there is a striking similarity between observations of real galaxies and data extracted from galaxies in a cosmological simulation, for the first time. An agreement like this was not seen in simulated galaxies until recently. This shows that state of the art cosmological simulations, are for the first time reproducing realistic galaxies, down to the finest of details.

On the basis of the agreement between the observed correlation and the theoretical correlation between SMBH mass and spiral arm pitch angle, Mutlu-Pakdil *et al* (2017) went on to look for a theoretical correlation between properties of the dark matter halo and the spiral arm pattern. Figure 6.9 shows the result of this investigation. Using the Illustris simulated galaxies, Mutlu-Pakdil *et al* (2017) demonstrated the existence of a theoretical correlation between (a) spiral arm pitch angle and the mass of dark matter in a galaxy halo (figure 6.9, left panel) and (b) spiral arm pitch angle and the total mass (dark plus luminous matter) in a galaxy halo (i.e., the halo mass, figure 6.9, right panel).

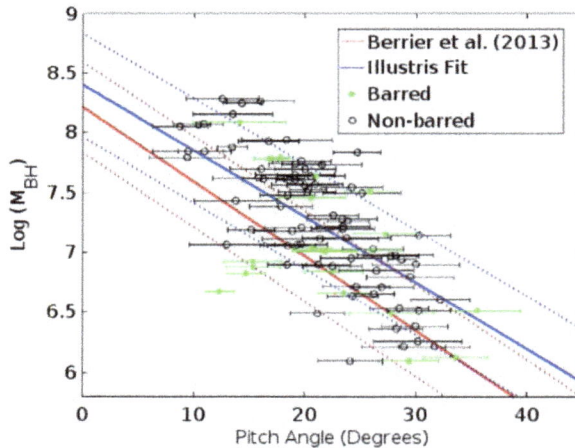

Figure 6.8. The Illustris prediction for the SMBH mass relative to spiral arm pitch angle is shown by the solid blue line, which represents the best fit to the data. The data points are based upon SMBH masses reported by the Illustris project and spiral arm pitch angles measured using images of galaxies in the Illustris simulation. The green stars represent barred galaxies, and the black circles represent non-barred galaxies. The dotted blue lines represent the scatter of ± 0.44 dex above and below the line. The observed relation from Berrier *et al* (2013) is represented by the solid red line with its scatter of ± 0.38 dex indicated by red dotted lines above and below the solid red line. Overall, the Illustris simulation reproduces the observed relation very well for disk galaxies.

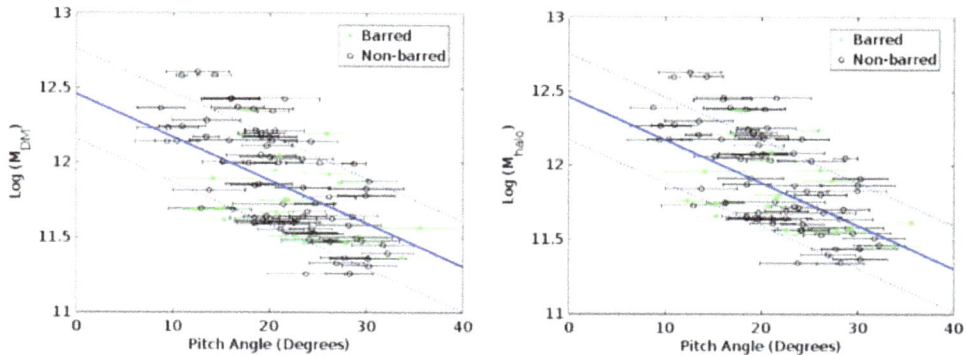

Figure 6.9. The Illustris predictions for the DM mass and halo mass relative to spiral arm pitch angle are shown by the solid blue lines, which represent the best fits to the data. Left panel: the relation between dark matter mass and pitch angle. Right panel: the relation between total halo mass and pitch angle. Barred galaxies are shown as green stars and non-barred galaxies are shown as black circles. The total RMS scatter is plotted using dotted blue lines above and below the best-fit lines, and in both cases the scatter is around ± 0.30 dex in the y-direction. Note that the Illustris predictions for these correlations are tighter than the one between pitch angle and the SMBH mass.

These results demonstrate, for the first time, a theoretical link between spiral arm pitch angle and the total mass of a disk galaxy. This goes beyond the prediction of density wave models, which said that there should be a relationship between pitch angle and the mass of the central bulge or central mass concentration. The cosmological simulations (in this case Illustris) seem to suggest that the total mass of the Galaxy (i.e., the total halo mass) is very tightly correlated with spiral arm pitch angle. Having said this, the results need to be confirmed observationally, but there is some intriguing observational evidence that may already be pointing in this direction, such as the relationship between spiral arm pitch angle and rotation curve shear shown in figure 6.2.

Suggested further reading

Berrier J C, Davis B L, Kennefick D, Kennefick J D, Seigar M S, Barrows R S, Hartley M, Shields D, Bentz M C and Lacy C H S 2013 *Astrophys. J.* **769** 132

Bertin G, Lin C C, Lowe S A and Thurstans 1989a *Astrophys. J.* **338** 78–103

Bertin G, Lin C C, Lowe S A and Thurstans 1989b *Astrophys. J.* **338** 104–20

Bertin G 1991 *Dynamics of Galaxies and Their Molecular Cloud Distribution* (IAU Symposium 146) ed F Combes and F Casoli (Dordrecht: Kluwer) pp 93–104

Bertin G 1993 *Publ. Astron. Soc. Pac.* **105** 640–3

Bertin 1996 *Astrophys. Space Sci. Libr.* **209** 227–42

Bertin G and Lin C C 1996 *Spiral Structure in Galaxies: A Density Wave Theory* (Cambridge, MA: MIT Press)

Block D L, Puerari I, Frogel J A, Eskridge P B, Stockton A and Fuchs B 1999 *Astrophys. Space Sci.* **269** 5–29

Bullock J S, Kolatt T S, Sigad Y, Somerville R S, Kravtsov A V, Klypin A A, Primack J R and Dekel A 2001 *Mon. Not. R. Astron. Soc.* **321** 559–75

Davis B L, Berrier J C, Shields D W, Kennefick D, Kennefick J, Seigar M S, Lacy C H S and
    Puerari I 2012 *Astrophys. J. Suppl.* **199** 33
Davis B L, Berrier J C, Johns L, Shields D W, Hartley M T, Kennefick D, Kennefick J, Seigar M
    S and Lacy C H S 2014 *Astrophys. J.* **789** 124
Davis B L, Kennefick D, Kennefick J, Westfall K B, Shields D W, Flatman R, Hartley M T,
    Berrier J C, Martinsson T P K and Swaters R A 2015 *Astrophys. J. Lett.* **802** L13
Elmegreen B G, Elmegreen D M and Hirst A 2004 *Astrophys. J.* **612** 191–201
Ferrarese L and Merritt D 2000 *Astrophys. J. Lett.* **539** L9–12
Fuchs B 1991 *Dynamics of Disk Galaxies* ed B Sundelius (Goteburg: Chalmers University of
    Technology) pp 359–63
Fuchs B 2000 Dynamics of galaxies: from the early Universe to the present (ed F Combes, G
    Mamon and V Charmandaris) *ASP Conf. Ser.* **197** 53–4
Gebhardt K *et al* 2000 *Astrophys. J. Lett.* **539** L13–6
Genel S, Vogelsberger M, Springel V, Sijacki D, Nelson D, Snyder G, Rodriguez-Gomez V,
    Torrey P and Hernquist L 2014 *Mon. Not. R. Astron. Soc.* **445** 175–200
Graham A W and Driver S P 2007 *Astrophys. J.* **655** 77–87
Graham A W, Driver S P, Allen P D and Liske J 2007 *Mon. Not. R. Astron. Soc.* **378** 198–210
Gultekin K, Richstone D O and Gebhardt K *et al* 2009 *Astrophys. J.* **698** 198–221
Haring N and Rix H-W 2004 *Astrophys. J. Lett.* **604** L89–92
Kennicutt R C Jr 1981 *Astron. J.* **86** 1847–58
Lin C C and Shu F H 1964 *Astrophys. J.* **140** 646–55
Magorrian J, Tremaine S and Rickstone D *et al* 1998 *Astron. J.* **115** 2285–305
Marconi A, Risaliti G, Gilli R, Hunt L K, Maiolino R and Salvati M 2004 *Mon. Not. R. Astron.
    Soc.* **351** 169–85
Mutlu-Pakdil B, Seigar M S and Davis B L 2016 *Astrophys. J.* **830** 117
Mutlu-Pakdil B, Seigar M S, Hewitt I B, Treuthardt P, Berrier J C and Koval L E 2017 *Mon. Not.
    R. Astron. Soc.* (submitted)
Navarro J F, Frenk C S and White S D M 1996 *Astrophys. J.* **462** 563–75
Navarro J F, Frenk C S and White S D M 1997 *Astrophys. J.* **490** 493–508
Roberts W W, Roberts M S and Shu F H 1975 *Astrophys. J.* **196** 381–405
Seigar M S and James P A 1998 *Mon. Not. R. Astron. Soc.* **299** 685–98
Seigar M S, Block D L and Puerari I 2004 *Astrophys. Space Sci. Lib.* **319** 155–63
Seigar M S, Block D L, Puerari I, Chorney N E and James P A 2005 *Mon. Not. R. Astron. Soc.*
    359 1065–76
Seigar M S, Bullock J S, Barth A J and Ho L C 2006 *Astrophys. J.* **645** 1012–23
Seigar M S 2008 *Publ. Astron. Soc. Pac.* **120** 945–51
Seigar M S, Barth A J and Bullock J S 2008a *Mon. Not. R. Astron. Soc.* **389** 1911–23
Seigar M S, Kennefick D, Kennefick J and Lacy C H S 2008b *Astrophys. J. Lett.* **678** L93–6
Seigar M S 2011 *ISRN Astron. Astrophys.* **2011** 725697
Seigar M S, Davis B L, Berrier J and Kennefick D 2014 *Astrophys. J.* **795** 90
van den Bergh S 1960 *Astrophys. J.* **131** 215–22
Vika M, Driver S P, Graham A W and Liske J 2009 *Mon. Not. R. Astron. Soc.* **400** 1451–60
Vogelsberger M, Genel S, Springer V, Torrey P, Sijacki D, Xu D, Snyder G, Nelson D and
    Hernquist L 2014 *Mon. Not. R. Astron. Soc.* **444** 1518–47
Weiner B J, Willmer C N A and Faber S M *et al* 2006 *Astrophys. J.* **653** 1027–48

Chapter 7

Concluding remarks

Of all of the proposed models that attempt to describe spiral structure in galaxies, modal theory appears to work best. It is important to point out that the modal theory as proposal by Guiseppe Bertin and collaborators, is essentially an extension of the original Lin–Shu Hypothesis from the 1960s. It is a form of density wave that consists of multiple spiral modes. Evidence in favor of the modal theory includes the following:

1. Fourier transformations of images of spiral galaxies reveal spiral modes of varying strength, with the even modes (particularly the $m = 2$ mode) dominating (e.g., Seigar and James 1998; Davis *et al* 2012).
2. The correlation between spiral arm pitch angle and shear (which is a proxy for the central mass concentration) is a prediction of the Lin–Shu hypothesis.
3. The regular nature and logarithmic nature of spiral arms in galaxies, as shown by Seigar and James (1998), Davis *et al* (2012) and Davis *et al* (2014).
4. The offset between spiral arms in different wavebands as one moves from optical to near-infrared wavebands (Seigar and James 1998; Pour Imani *et al* 2016).
5. A relation between spiral arm strength (as measured in near-infrared K-band images) and star formation rates (as measured from narrow-band Hα images), as predicted by the large-scale shock scenario, which is a result of the Lin–Shu hypothesis (Seigar and James 2002).

The above list includes significant evidence in favor of the modal theory of spiral structure, but it is by no means exhaustive. For now we will discuss, in a little more detail the idea that the correlation between central mass concentration and spiral arm pitch angle supports modal theories of spiral structure.

So how is this connection between central mass concentration and spiral arm pitch angle consistent with other relations, such as those between spiral arm pitch

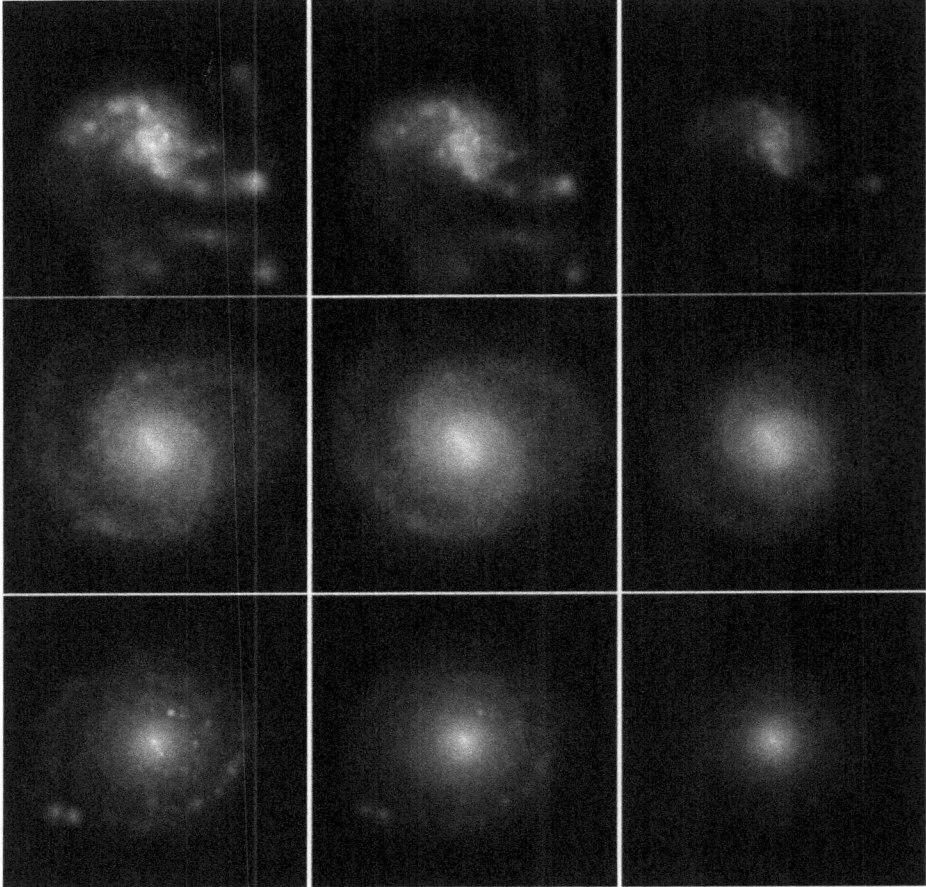

Figure 7.1. Example images in the **B** (blue), **R** (red), and **K** (near-infrared) filters for three $z = 0$ Illustris galaxies are shown from right to left, respectively. They are arranged by increasing stellar mass from $M_* \sim 10^{10}$ solar masses (bottom) to $M_* \sim 10^{11}$ solar masses (top). Figure is from Mutlu-Pakdil *et al* (2017).

angle and the properties (or mass) of the dark matter halo in disk galaxies? This can be answered in several ways:

1. Firstly, the mass of the bulge (or spheroidal) component in galaxies correlates strongly with the dark matter halo mass (e.g., Grand *et al* 2017) both in simulations and observations.
2. The galaxy total stellar mass (i.e., the bulge plus disk stellar mass) also seems to correlate with the dark matter halo mass, particularly in simulations (e.g., Mutlu-Pakdil *et al* 2017 and the results of cosmological simulations such as Illustris as shown in figure 7.1).
3. The bulge stellar mass correlates well with both SMBH mass (e.g., Haring and Rix 2004) and with pitch angle (Davis *et al* 2014).

Given these three points above, it follows naturally that spiral arm pitch angle would correlate with central mass concentration (or bulge mass) because simulations and observations both suggest that these parameters may actually be governed by more global properties, such as the dark matter halo mass.

So what can we take away from this? Firstly, spiral structure really does appear to be the result of a density wave, and the modal theory seems to describe spiral structure the best. In this model, a density wave arises as a natural gravitational instability in a disk of stars. Most (if not all) of the evidence appears to be pointing in this direction, and it really appears that the mystery behind spiral structure in galaxies has now been solved. Nevertheless, spiral structure is an important morphological feature in disk galaxies. In the last decade it has become evident that parameters such as spiral arm pitch angle are very strongly correlated with other galaxy properties such as bulge mass, supermassive black hole mass, and dark matter halo mass. If we use spiral arm morphology as a proxy for these mass indicators in galaxies, we have a very strong tool for looking at mass buildup in galaxies over cosmological times, which in turn will help astronomers narrow down our current cosmological models. This will be the main use for studies of spiral arm morphology over the coming decade.

Suggested further reading

Davis B L, Berrier J C, Shields D W, Kennefick D, Kennefick J, Seigar M S, Lacy C H S and Puerari I 2012 *Astrophys. J. Suppl.* **199** 33

Davis B L, Berrier J C, Johns L, Shields D W, Hartley M T, Kennefick D, Kennefick J, Seigar M S and Lacy C H S 2014 *Astrophys. J.* **789** 124

Grand R J J, Gomez F A, Marinacci F, Pakmor R, Springer V, Campbell D J R, Frenk C S, Jenkins A and White S D M 2017 *Mon. Not. R. Astron. Soc.* **467** 179–207

Haring N and Rix H-W 2004 *Astrophys. J. Lett.* **604** 89–92

Mutlu-Pakdil B, Seigar M S, Hewitt I B, Treuthardt P, Berrier J C and Koval L E 2017 *Mon. Not. R. Astron. Soc.* (submitted)

Pour Imani H, Kennefick D, Kennefick J, Davis B L, Shields D W and Shameer Abdeen M 2016 *Astrophys. J. Lett.* **827** L2

Seigar M S and James P A 1998 *Mon. Not. R. Astron. Soc.* **299** 685–98

Seigar M S and James P A 2002 *Mon. Not. R. Astron. Soc.* **337** 1113–7

www.ingramcontent.com/pod-product-compliance
Lightning Source LLC
Chambersburg PA
CBHW082111210326
41599CB00033B/6662